Upper Cambrian conodonts fror

KLAUS J. MÜLLER and INGELORE HINZ

Müller, Klaus J. & Hinz, Ingelore 1991 02 15: Upper Cambrian conodonts from Sweden. *Fossils and Strata*, No. 28, pp. 1–153. Oslo. ISSN 0300-9491. ISBN 82-00-37475-0.

More than 600 samples from Västergötland and the isle of Öland, covering the *Agnostus pisiformis* up to the *Peltura scarabaeoides* zone, have been examined for conodonts. The fauna comprises 73 species of 20 genera. It yields highly differentiated elements such as the ramiform *Serratocambria* and the pectiniform *Cambropustula*, the latter being the oldest known conodontophorid. The genera *Furnishina* and *Westergaardodina* have not only the greatest number of species but also of individual representatives. The genus *Proscandodus* has been abolished and the according taxa reclassified. The occurrence of individual conodont taxa has been correlated with the trilobite-based zonal stratigraphy. The fauna is being discussed under various geological and biological aspects, supplemented by detailed systematic studies including instraspecific variability as well as morphogenesis during growth of the animal. Newly established genera are: *Bengtsonella, Cambropustula, Gumella, Serratocambria* and *Trolmenia*. New species are: *Bengtsonella triangularis, Cambropustula kinnekullensis, Furnishina curvata, F. gladiata, F. gossmannae, F. kleithria, F. kranzae, F. mira, F. ovata, F. sinuata, F. vasmerae, Gumella cuneata, Muellerodus guttulus, M. subsymmetricus, Nogamiconus falcifer, Phakelodus simplex, Problematoconites angustus, P. asymmetricus, Prosagittodontus minimus, Serratocambria minuta, Trolmenia acies, Westergaardodina ahlbergi, W. auris, W. behrae, W. calix, W. communis, W. concamerata, W. curvata, W. excentrica, W. latidentata, W. ligula, W. polymorpha, W. procera, W. prominens.* □ *Conodonts, systematics, Cambrian stratigraphy, anthraconite, orsten, Sweden, N5615 N5836 E1630 E1225.*

Klaus J. Müller & Ingelore Hinz, Institut für Paläontologie, Rheinische Friedrich-Wilhelms-Universität, Nußallee 8, D-5300 Bonn 1, Federal Republic of Germany; 1990 02 19 (revised 1990 08 23).

Contents

Introduction

The material described herein has been collected from various outcrops in Västergötland and from the isle of Öland (Fig. 1). For this region it represents the entire stratigraphic series between the *Agnostus pisiformis* and the *Peltura scarabaeoides* zones. The study is a continuation of the first systematic description of Cambrian conodonts by Müller (1959), the European material in that study mainly coming from Sweden. At that time, however, working conditions were limited compared with today's space, equipment and technical help. Because additional localities have been included in the study and many of the previous sites have been resampled in more detail, a considerably improved investigation could be achieved. This has led to a better understanding of these early sclerites during a time interval that is seminal for many of the important evolutionary lineages present in the Lower Ordovician.

Intensified work on this material dates back more than a decade and has been combined with the search for small arthropods having their chitinous body and appendages preserved (Müller 1979; Müller & Walossek 1988). This has resulted in one of the most extensive conodont collections from the Cambrian. In general, conodonts of this age are sparsely scattered. For example, Druce & Jones (1971, p. 25) processed more than 1400 kg of limestones from which they recovered only 5000 specimens.

The advantages of methodical studies based on large faunal associations are as follows:

(a) Many additional taxa have been encountered, contradicting Bengtson's previous conclusion that Cambrian conodonts have low diversity as well as simple morphology and structure (Bengtson 1983b).

(b) The stratigraphic range of individual species can be defined more precisely.

(c) The chance to recover clusters is increased.

(d) Material suitable for histological studies can be selected from a larger stock.

(e) Morphological features can be recognized and compared more comprehensively, improving our knowledge about the morphogenesis during ontogeny, variation etc. Convergent aspects between Cambrian and younger conodonts include general characters of the gross morphology. Examples, at least on the generic level, of similar outer morphology but a different internal structure are *Prooneotodus/Oneotodus*, *Prosagittodontus/Sagittodontus*, *Cambropustula/Polonodus*, as well as *Gapparodus* and *Gumella/Panderodus*. A recurved cusp with uplifted tip, for example, is frequently observed on some *Furnishina* species and *Muellerodus*.

(f) Large associations demonstrate that the reconstruction of natural associations based on isolated specimens is extremely complicated due to mixture of the taxa before final sedimentation, especially in a more stratigraphically condensed sequence.

(g) This study documents the significance of sample size to data on the distribution of individual taxa. For example, the previous non-record of *Proconodontus* in Sweden, simply a collecting bias, made Miller (1984) assume temperature-controlled distributional differences for that genus. As the majority of published occurrences is based on less material than even the 2870 specimens of Müller's (1959) original study, they obviously can have only limited value for long distance correlation, regional differences, etc.

In addition, this study focussed on the correlation of conodont index species with the well-established trilobite zones. Depending on the preservation potential it might be easier to obtain identifiable conodonts than trilobites. But conodont taxa turned out to be less precise time markers than trilobite ones. Nevertheless, a number of species are useful to indicate stratigraphic position within the series.

The Upper Cambrian conodont fauna is characterized by the predominance of para- and protoconodonts which gave rise to a large variety of simple cone elements. The genera *Furnishina* and *Westergaardodina* are the most frequent ones and are represented by a great number of species. Until now, euconodonts had been reported only from near the Cambrian–Ordovician boundary. But the highly developed genus *Cambropustula* occurs already in

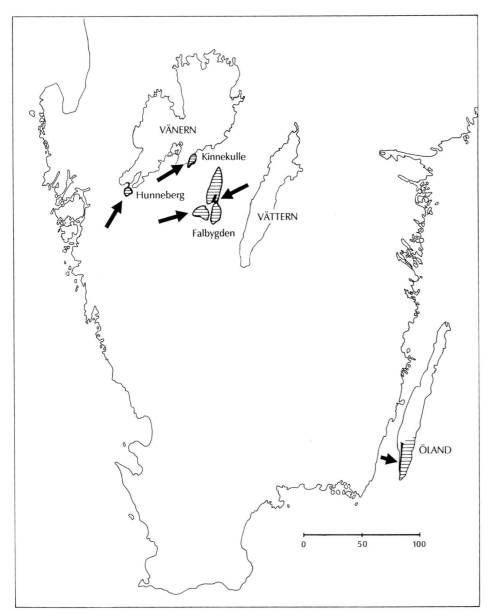

Fig. 1. General view of south and middle Sweden with the sampled regions marked by arrows.

the lowermost Upper Cambrian. *Cambropustula* and the age-equivalent *Acodus cambricus* are the oldest known representatives of the Conodontophorida.

History of study

The record of Cambrian conodonts starts with Wiman's discovery of problematic fossils of unidentified age from Sweden (Wiman 1893). Westergård (1953) described and illustrated a variety of such bi- and tricuspidate specimens as 'Problematicum I' from South and Middle Sweden with a recorded age of late Middle to late Late Cambrian.

Exactly a century after the publication of Pander's fundamental discovery of conodonts, Müller (1956) initiated the documentation of Cambrian forms, and in 1959 he presented the first systematic study of Middle but predominantly Upper Cambrian taxa from Northern Europe and the United States. 'Problematicum I' of Westergård was

included under the generic name *Westergaardodina*. Initially, these forms were rejected as conodonts 'at least strictly speaking' (Lindström 1964) before they received acceptance as true representatives of the group by Clark & Robison (1969), Druce & Jones (1971), Szaniawski (1971), Miller (1981), etc.

Unexpectedly low activity in this particular field, except for Grant (1965), Poulsen (1966), Nogami (1966, 1967), and Clark & Robison (1969) continued through most of the 1960's. In 1969, Miller published the first results of his investigation of the Cambrian–Ordovician boundary in North America; he discovered completely new associations which turned out to be important index fossils. With the introduction of stratigraphic aspects, Miller set a trend for many conodont workers to focus on that particular time interval. As a consequence, these faunas are much better studied than their predecessors. As they seem to be less provincial than trilobites or graptolites, respectively, the International Working Group for the Cambro–Ordovician

Boundary (IWGCOB) decided in 1985 to choose cono-donts as the main fossil group for a redefinition of the base of the Ordovician. Presently that boundary is defined differently, and is at different levels, in separate regions.

During the last 20 years Cambrian conodont research has covered a number of different fields such as stratigraphy, phylogeny and evolution, morphology, histology and function, systematic position, facies, provincialism, temperature control, geochemistry and chemoevolution. Some of these aspects require new methods, techniques and highly specialised equipment.

Stratigraphy. – Most effort in Cambrian conodont research has so far been devoted to stratigraphy. Particular attention has been paid to the Cambrian–Ordovician boundary on all continents except Africa.

The faunas of the United States have been studied by Miller (1969, 1970, 1978, 1980, 1984, 1987, 1988a), Kurtz (1976), Landing (1974, 1979, 1983), Miller, R.H. *et al.* (1981), Taylor, Repetski & Sprinkle (1981), Taylor & Landing (1982), Miller *et al.* (1982), Dutro *et al.* (1984), Hintze, Taylor & Miller (1988) and Orndorff (1988).

Canadian occurrences have been covered by Derby, Lane & Norford (1972), Tipnis, Chatterton & Ludvigsen (1978), Fåhraeus & Nowlan (1978), Landing, Taylor & Erdtmann (1978), Landing (1980), Landing, Ludvigsen & von Bitter (1980), Fortey, Landing & Skevington (1982), and Nowlan (1985).

The East Greenland boundary interval was subject to a report by Miller & Kurtz (1979).

Scandinavia was partly covered by Müller (1959) and Bruton, Koch & Repetski (1988).

Equivalent faunas in the USSR were studied by Abaimova & Markov (1977), Abaimova (1978), Apollonov & Chugaeva (1982), Dubinina (1982), Apollonov, Chugaeva & Dubinina (1984), Kaljo *et al.* (1986), Heinsalu *et al.* (1987), and Viira, Sergeeva & Popov (1987).

In China, numerous sections in different parts of the country have been investigated by Nogami (1966, 1967), An & Yang (1980), Chang, Chu & Lin (1980), An (1982, 1987), Lu & Mu (1980), An *et al.* (1983), An, Du & Gao (1985), Dong (1984, 1986, 1987, 1988), Wang (1985a, b), Wang & Li (1986), Chen & Gong (1986), Zhao (1986), Chen, Zhang & Yu (1986), and Ding, Bao & Cao (1987).

South Korean occurrences have been studied by Lee (1975) and Lee & Lee (1988) Indian sections by Azmi, Yoshi & Juyal (1981), Azmi (1983a, b), Azmi & Pancholi (1983). Müller (1973) described boundary interval faunas from Iran, as did Özgül & Gedik (1973) from Turkey.

In Australia, the first Cambrian conodonts were recorded by Jones (1961). A conodont-based stratigraphy has been worked out by Druce & Jones (1968), Jones (1971), Jones, Shergold & Druce (1971), and Druce, Shergold & Radtke (1982).

Buggisch (1982), Burrett & Findlay (1984), Wright, Ross & Repetski (1984) and Buggisch & Repetski (1987) noted findings in Antarctica.

Heredia & Bordonaro (1988) recorded conodonts from Argentina. Faunas from Mexico were reported by Miller, Robison & Clark (1974).

Lower and Middle Cambrian conodonts, as well as lower Upper Cambrian ones, have been less studied, partly due to their rarity. This field, which undoubtedly is very important for the development of this group as a whole, has been partly covered by Müller (1959: northern Europe, USA), Nogami (1966: China), Clark & Robison (1969: USA), Meshkova (1969: USSR), Szaniawski (1971: Poland), Missarzhevsky (1973: USSR) Landing (1974: USA), Miller, R.H. & Paden (1976: USA) MacKinnon (1976: New Zealand), Abaimova (1978: USSR), Bednarczyk (1979: Poland), Bhatt (1980: India), Missarzhevsky & Mambetov (1981: USSR), An (1982: China), Azmi & Pancholi (1983: India), Brasier (1984: England), Dutro *et al.* (1984: USA), Jiang *et al.* (1986: China), Bischoff & Prendergast (1987: Australia), as well as by Hinz (1987: England).

Function. – Based on their shape, Pander (1856) was convinced that conodonts functioned as teeth; he therefore named the whole group 'conodonts'. This view was generally accepted until the discovery of the centrifugal mode of growth (Hass 1941; Gross 1954), which requires enclosure by a secreting tissue. As a consequence, the sclerites were interpreted as internal supporting organs. Lindström (1964, 1974) and Conway Morris (1976, 1978), e.g., proposed a function as lophophorate-supporting structures. By contrast, Carls (1977) pointed out that sharp edges favour an external function rather than an endoskeletal one. The morphological comparisons with known teeth were further developed by Jeppsson (1979). Carls (1977) assumed that the elements could be expelled and substituted, which would lead to an unbalanced numerical record in the samples with regard to the known apparatuses.

The discovery of the paraconodont (Müller & Nogami 1971, 1972a, b) and protoconodont (Bengtson 1976, 1983b) modes of growth revealed a certain resemblance to the centripetal structure of teeth. Accordingly, these sclerites have been regarded as external organs.

With findings of the protoconodont *Phakelodus* clusters (Miller & Rushton 1973), functional interpretation has stressed the external operation of these sclerites as grasping spines similar to those of the Recent chaetognath *Sagitta* (Müller & Andres 1976; Landing 1977; Szaniawski 1980b, 1982, 1983, 1987; Repetski & Szaniawski 1981; Andres 1981; Bengtson 1983a and Sweet 1985).

Covering the present state of knowledge, Bengtson (1976, 1983a) proposed an external function, not only for proto- and paraconodonts, but also for conodontophorids, assuming that they had alternating growth and functional phases.

Evolution. – Clark & Miller (1969) noted a 'mineralogical evolution from calcium phosphate and a considerable amount of organic matter in the oldest specimens to calcium phosphate with little ... organic material in latest Cambrian specimens' (see also Repetski & Szaniawski 1981). Miller (1980) implied that colour differences between para- and associated euconodonts are due to differences in the organic content. Based on morphological details several lineages have been proposed: *Proconodontus–Cordylodus* (Clark & Miller 1969), *Teridontus–Hirsutodontus* and *Eoconodontus–Cordylodus* (Miller 1980).

Already in 1976 Miller recorded a primitive new euconodont genus, the presumed missing link between the paraconodont *Prooneotodus rotundatus* and the conodontophorid *Proconodontus tenuiserratus*, which acquired a crown and gradually a finely serrated posterior keel. Due to morphological similarities, several authors believe in a polyphyletic origin of the euconodonts or at least cannot exclude it (Miller 1976, 1984; Chen & Gong 1986; Szaniawski & Bengtson 1988). Also Andres (1988) discussed the relationship between proto-, para- and euconodonts.

The genus *Chosonodina* was considered as being derived from *Westergaardodina* by Müller (1964) and Druce & Jones (1971). Miller (1984) noted the presence of white matter and consequently referred *Chosonodina* to the conodontophorids, postulating a transition from para- to euconodonts in this lineage. However, the critical difference between westergaardodinids and euconodonts is the growth direction which has not been yet investigated for *Chosonodina*.

According to Bengtson (1983b) transitions between proto- and paraconodonts are unknown and more hypothetic than that between para- and euconodonts. In any case, such transitions are difficult to prove as the lower rim of protoconodonts is extremely thin and thus even an upturning lamella is hardly ever preserved.

Open questions. – Among the many problems which still await their tackling, two fields seem to be of particular interest: geochemistry and histology. The geochemistry of conodonts is a subject barely touched. Miller (1987) pointed out that trace elements in the apatite of conodonts and inarticulate brachiopods display correlative variations. Further studies have to prove if such differences are of general significance. Concerning their internal structure, many of the Cambrian conodont taxa have not been examined yet, but such a study by us is underway. It may lead to a better understanding of the systematic relationships.

Material and methods

The fossiliferous anthraconite (*orsten*) has been collected from zones I–V of the Upper Cambrian. It derives from numerous outcrops at the Kinnekulle (Fig. 2), in Falbygden (Fig. 3) and at the Hunneberg in Västergötland as well as from the isle of Öland (Fig. 4). The large number of outcrops is due to mining activities for alum shale during the 18th and 19th century. Some erratic boulders from Northern Germany also have been added to the collections. The localities are listed in Appendix 1.

Within the *orsten*, six different lithologies, independent of the megafossil content, have been collected; the individual types are described below. They are not correlatable to specific zones but occur throughout the stratigraphic sequence. Often they are in close contact with each other even at very small scale, so that they could not be collected separately. For this reason, a relation between lithology and conodont content could not be observed. In larger outcrops, several samples have been taken from a single bed

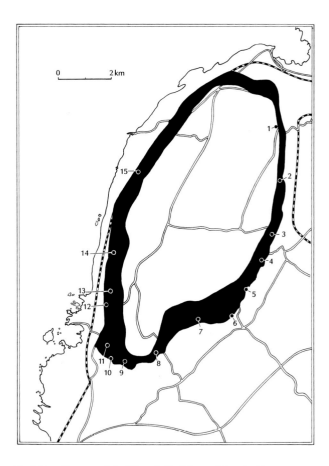

Fig. 2. Locality map of the Kinnekulle, Västergötland, with outcrops of the Upper Cambrian Alum Shale (black). 1: Gössäter. 2: Toreborg. 3: Österplana. 4: Haggården–Marieberg. 5: Ödbogården. 6: Brattefors. 7: Sandtorp. 8: Stubbegården. 9: Gum. 10: Ekebacka. 11: Backeborg. 12: Klippan. 13: Kakeled. 14: Pusabäcken. 15: Trolmen.

or nodules of the same zone. Even within very short distances, the microfossil composition may vary in abundance and preservation.

Type A: Black to grey, micritic to sparry, thin-layered limestones. The carbonaceous content is fairly high (stinkstone). In most cases the fossils, predominantly exuviae of trilobites, are rock-forming. There is some compaction documented by partly compressed megafossils. Sorting is quite common.

Type B: Beige, mostly sparry limestones. The trilobites are as abundant as in type A and similarly preserved. Some rocks are spotted with phosphatic precipitation. This type may occur either in banks or together with type A as thin alternating layers.

Type C: Black, fine-grained limestones with a high organic content. Quite commonly, the original micritic matrix has been recrystallized. The fossil content is rather low; less than 5%. This type may occur either in layers or as inclusions in one of the other types.

Type D: Light grey, micritic to sparry limestones. Phosphatic precipitation occurs in dark spots. The coarsely hashed megafossil content occupies approximately 50%.

Type E: Black, micritic, thin-layered limestones with fossil hash. In some cases the latter appears to be rock-forming.

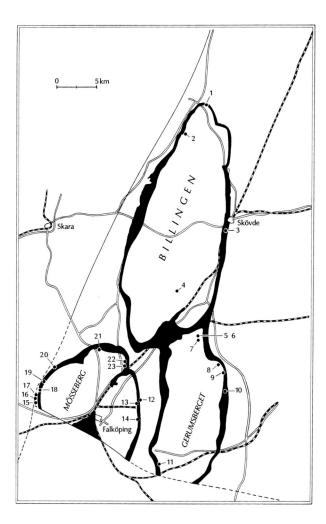

Fig. 3. Locality map of Falbygden, Västergötland, with outcrops of the Upper Cambrian Alum Shale (black). 1: St. Stolan. 2: Karlsfors. 3: Karlsro. 4: Ranstadsverket. 5: Nya Dala. 6: Stenstorp–Dala. 7: Stenåsen. 8: Smedsgården–Stutagården. 9: Ekedalen. 10: Ödegården. 11: Milltorp. 12, 13: Uddagården. 14: Djupadalen. 15: Nästegården. 16: Ekeberget. 17: Ledsgården. 18: Skår. 19: Kleva. 20: St. Backor. 21: Gudhem. 22: Rörsberga. 23: Tomten.

Type F: Whitish to grey weathered limestones with a high content of fine fossil hash. Trilobites are the prevailing elements.

The sampling focussed on limestones with rich and well-preserved microfossils. Accordingly, a number of criteria determined the mode of sampling as follows: (1) Pure limestones leaving only small etched residues were preferred. Such rocks are usually light beige to grey coloured. On the other hand, deeply black limestones, rich in organic matter, are less suitable. (2) The limestones should be fresh and unweathered. (3) Larger veins of secondary calcite were avoided. (4) Rocks exposing phosphatocopids had been preferably collected. But without a hand lens, these carapaces are extremely difficult to observe. In addition, such material is rare and generally requires intensive search. (5) According to previous experience, small and evenly rounded nodules with smooth surface are most productive. As they are not common, all of them were collected.

It is clear that the above mentioned criteria prevented sampling in a vertical section with detailed measurements

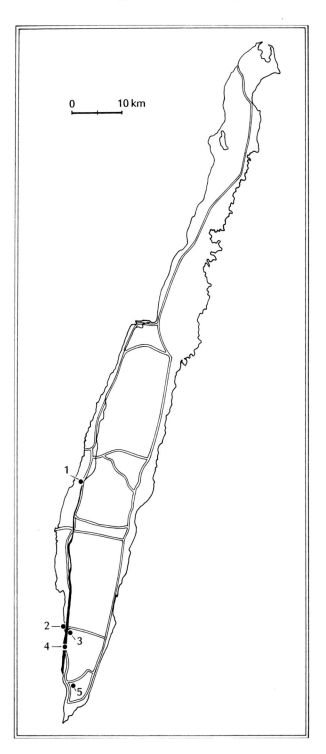

Fig. 4. Locality map of the Isle of Öland with outcrops of the Upper Cambrian Alum Shale (black). 1: Eriksöre. 2: Degerhamn. 3: S. Möckleby. 4: Mörbylilla. 5: Grönhögen.

in the various outcrops. The age of the individual samples was determined on the basis of trilobites. This turned out to be sufficiently precise because the conodonts are long-ranging and unsuitable to indicate zones or subzones in the succession.

The volume of the samples varied between several hundred grams and ten kilograms depending on geological occurrence, quality of preservation and faunal content. In many cases it was limited by the size of the individual nodules.

In order to obtain soft-bodied fossils, the sampling concentrated on certain types which mainly came from zones I and V. This resulted in a somewhat unbalanced stratigraphic representation. Furthermore, the frequency of individual species may be influenced by uneven concentration in the same layer. In larger outcrops, several samples have been taken from a single bed or nodules of the same horizon. Even within very short distances, the microfossil composition may vary in abundance and preservation. Locally, mass occurrences of conodonts can be observed in the field.

More than 1200 kg rock material of 550 productive samples was processed by routine etching with 10–15% acetic acid. Despite its obvious fossil content, the alum shale was not included in this study as it is unsuitable for the applied preparation technique. Due to the availability of large individual samples, the addition of a buffer to the acid was unnecessary. The calcium-acetate generated by dissolution of the first portion of limestone served as buffer (Müller 1985). The application of screens during the etching process helped to preserve fragile clusters with up to 26 sclerites. Heavy-liquid techniques, as well as interfacial and magnetic separation, were avoided because the soft-bodied arthropods would not have survived such a rough procedure. Without further concentration, the residues were picked under stereo-lenses, mounted on SEM stubs with double-sided sellotape or a high-vacuum wax and finally coated with carbon and/or gold.

Morphological characters have been documented by CamScan II photography; a total of 996 illustrations is presented. To maintain size relations among the individual sclerites, different magnifications within a single species were sometimes required. Extremes towards a lower and/or upper size limit could also be avoided. In a few cases, even various views of the same specimen have been reproduced at different magnifications to avoid large size differences due to distortion.

The synonymy lists are selective. Where critical evidence is missing from published illustrations or text, the references have not been included. Based on our extensive material, individual taxa could be defined more precisely than in Müller 1959. This has led to new species concepts that involve revisions of some previous determinations. Much emphasis was laid upon the recognition of changes in morphogenesis and intraspecific variation (based mainly on differences between specimens of equal size). Morphological change may prevent the recognition of large and small elements as belonging to the same species without knowledge of the intermediate stages. The degree of change during growth can be quite variable for individual taxa.

For the reconstruction of apparatuses, numerical proportions of disjunct elements have not been elaborated for the following reasons: (1) We consider an ontogenetic increase of sclerites likely. In this context it should be kept in mind that very small stages might be too undifferentiated for a specific determination. (2) Loss and replacement of conodont elements, as postulated by Carls (1977), cannot be ruled out for our material. Therefore relative frequency studies are of only limited value in this respect.

Acknowledgements

We are deeply indebted to Per Ahlberg, Lund, for kindly aiding us in the determination of trilobites for the stratigraphic zonation. Also we would like to thank Gunnar Henningsmön, Oslo, for help in this respect. Thanks are due to John Repetski, Washington, for carefully reviewing the manuscript. Valuable technical help was contributed by Ms. Andrea Behr, Mrs. Annemarie Gossmann, Ms. Angela Hille, Ms. Dorothea Kranz, Mr. Georg Oleschinski, Mrs. Marianne Vasmer-Ehses and others. We greatly appreciate the financial support by the *Deutsche Forschungsgemeinschaft*. The publication was supported by a grant from the Swedish Natural Science Research Council.

Regional setting and stratigraphy

On the Baltoscandian Platform, the eastern foreland of the Scandinavian Caledonides, Cambrian shallow-water sediments were deposited under generally stable conditions. After a peneplanation in the Precambrian, the whole area faced repeated transgressions and regressions. The regionally uniform sediments have been influenced by minor vertical movements only. The Lower Cambrian consists of sandy shallow-water sediments. By contrast, the Middle Cambrian through Tremadocian is represented by dark bituminous shales with increasingly common intercalated stinkstones in the Upper Cambrian, the so-called *orsten*. The entire succession was deposited in a large, almost closed basin and is known as the Alum Shale Formation. Except for the Oslo and Skåne region it has a thickness of only 20–30 m but a great areal extent, covering also the Baltic Sea and large parts of Poland, as has been proved from core samples there. The series is considerably condensed. Even within a thickness of 10 cm, up to three subzones have been observed. Differences in the development, i.e. thickness of individual beds and local hiata, are common.

An interesting and important feature is the remarkable geochemistry of the alum shale. It is rich in trace elements, particularly in uranium and vanadium. The maximum content has been recorded from the *Peltura* zones of the Upper Cambrian which may even exceed the most uraniferous black shales of the Chattanooga Shale in North America (Bergström & Gee 1985). Apparently this concentration coincides with the the most fossiliferous deposits of the whole series.

The fauna of this particular time span, with its swimming organisms, differs from the rich benthos of the Middle Cambrian and Tremadocian. Environmental conditions with reduced oxygenisation led to a somewhat restricted benthic life represented mainly by small arthropods. Large trilobites and brachiopods are known only from light-coloured sediments that had obviously better aeration.

Nevertheless, it has to be kept in mind that the abundant but faunally restricted occurrence represents a regional feature. This assumption is strengthened by the non-record

Table 1. Trilobite zones of the Upper Cambrian of Sweden. After Westergård 1922 and Henningsmoen 1957.

	Peltura scarabaeoides	Vc
V	*Peltura minor*	Vb
	Protopeltura praecursor	Va
IV	*Leptoplastus*	
III	*Parabolina spinulosa*	
II	*Olenus*	
I	*Agnostus pisiformis*	

Table 2. Range chart of the conodont taxa. Open symbols refer to uncertain stratigraphic position.

	I	II	III	IV	Va	Vb	Vc	V und.
Bengtsonella triangularis						●	●	●
Furnishina alata	●							
Furnishina asymmetrica	●	●	●	●		●	●	
Furnishina bicarinata	●	●	●	●	●	●	●	●
Furnishina curvata			●	●	●	●	●	●
Furnishina furnishi			●	●	●	●	●	●
Furnishina gladiata	●	●						
Furnishina gossmannae			●	●	●	●	●	●
Furnishina kleithria	●						●	●
Furnishina kranzae		●						
Furnishina longibasis		●						
Furnishina mira	●	●				●	●	●
Furnishina ovata					●	●	●	●
Furnishina polonica	●	●	●	●				
Furnishina primitiva			●			●	●	●
Furnishina quadrata			●	●		●	●	●
Furnishina rara	●	●	●	●	○	●		●
Furnishina sinuata			●	●		●	●	●
Furnishina tortilis	●	●	●	●		●	●	●
Furnishina vasmerae	●	●						
Gapparodus bisulcatus	●	●					●	●
Gumella cuneata	●	●					●	
Hertzina elongata	●	●	●	●	●	●	●	●
Muellerodus cambricus	●	●	●	●		●		●
Muellerodus guttulus	●	●				●		●
Muellerodus? oelandicus	●	●	●	●	●	●	●	●
Muellerodus pomeranensis	●	●	●	●	○	●	●	●

	I	II	III	IV	Va	Vb	Vc	V und.
Muellerodus subsymmetricus	●	●						
Nogamiconus falcifer	●	●	●		○	●	●	●
Nogamiconus sinensis	●		●	●				
Phakelodus elongatus	●	●	●			●	●	●
Phakelodus simplex	●	●				●	●	●
Phakelodus tenuis	●	●	●	●	○	●	●	●
Proacodus obliquus	●	●	●			●	●	●
Proacodus pulcherus		●	●		○	●	●	●
Problematoconites angustus		●	●		○	●	●	●
Problematoconites asymmetricus		●			○	●		●
Problematoconites perforatus	●	●	●	●	●	●	●	●
Prooneotodus gallatini						●	●	●
Prosagittodontus dahlmani	●	●				●	●	●
Prosagittodontus minimus						●	●	●
Serratocambria minuta		●				●	●	●
Trolmenia acies						●	●	●
Westergaardodina ahlbergi	●	●						
Westergaardodina amplicava		●		●	●	●	●	●
Westergaardodina auris	●							
Westergaardodina behrae					○	●	●	●
Westergaardodina bicuspidata	●	●	●	●	○	●	●	●
Westergaardodina bohlini			●	●		●	●	●
Westergaardodina calix					○	●	●	●
Westergaardodina communis	●	●						
Westergaardodina concamerata					○	●	●	●
Westergaardodina curvata						●	●	●
Westergaardodina excentrica	●	●						
Westergaardodina latidentata						●	●	●
Westergaardodina ligula					●	●	●	●
Westergaardodina matsushitai	●	●						
Westergaardodina microdentata		●	●	●	○	●	●	●
Westergaardodina moessebergensis	●	●						
Westergaardodina nogamii	●	●	●					
Westergaardodina obliqua	●							
Westergaardodina polymorpha	●	●		○		●	●	●
Westergaardodina procera	●					●	●	●
Westergaardodina prominens		●			○	●	●	●
Westergaardodina quadrata	●	●						
Westergaardodina tricuspidata		●	●					●
Westergaardodina wimani	●	●						
Acodus cambricus	●	●	●			●	●	●
Cambropustula kinnekullensis	●	●						
Coelocerodontus bicostatus	●					●	●	●
Cordylodus primitivus							●	●
Proconodontus muelleri							●	●
Proconodontus serratus							●	●

of molluscs, generally a common group in Cambrian strata. Apart from that, locally concentrated occurrence is a common character of some faunal elements. For example, although trilobites have been studied extensively for more than 200 years by various authors, Henningsmoen (personal communication, 1982) made the first discovery of *Acrocephalites stenometopus* (Angelin) from Västergötland in one of our samples. On the other hand, individual taxa like *Phakelodus tenuis* may be the only conodont component in a sample, while in other ones of the same age different species prevail. Also, other mass occurrences of single species have been observed, e.g., *Orusia lenticularis* or *Agnostus pisiformis*, with only a limited variety of associated elements. These special conditions are of great importance for comparison with other regions. Compositional differences in beds of similar age cannot be simply referred to climate or faunal provincialism etc. It is likely that fossils which are regionally useful as time markers may occur in different levels elsewhere. The occurrence of endemic forms should also be considered.

Based on trilobites, the Upper Cambrian was divided into six successive zones by Westergård (1922) and Henningsmoen (1957) (Table 1). Their results served as basis for this study. However, Schrank (1973), who described the trilobites of Upper Cambrian stinkstone boulders from Northern Germany and Poland, came to the conclusion that the 'Scandinavian trilobite zones with their subdivision worked out by Henningsmoen are not always distinctly separated. Species of adjacent zones may occur together in the same rock' (translated from German).

With regard to conodonts, the *orsten* turned out to be very productive in parts, but a complete sequence between the *Agnostus pisiformis* and the *Peltura scarabaeoides* zones was not obtained. The age determination of the individual samples is based on their trilobite content, but for lack of suitable material it was sometimes difficult to identify the guide fossils for the respective subzones. A correlation between trilobite zones and conodont ranges is given in Table 2.

The conodont distribution is currently not considered to be of orthostratigraphic value, because the succession is recognized only for the study area. The first appearance of certain species of *Acodus*, *Coelocerodontus*, *Furnishina* and *Westergaardodina* is not consistent with other areas (compare, e.g., Nogami 1966; Druce & Jones 1971 and van Wamel 1974). A number of species which are widely distributed in both the lower and uppermost zones in Sweden have not been recovered from trilobite zones III and IV there. Possible reasons are discussed in the chapters *Material and methods* and *Factors limiting species distribution*. Only further study in other regions and/or paleoclimates etc., can establish wider orthostratigraphic value.

Worldwide comparison

Most previous Upper Cambrian conodont studies concern the Cambrian–Ordovician boundary, with faunas not directly comparable with the earlier ones studied here. Only An *et al.* (1983) established zones prior to the *Proconodontus* Zone in North China and adjacent regions, but this zonation has not been confirmed elsewhere.

Of 80 species in the present study, 45 have been recorded also from localities outside the Alum Shale Basin. They indicate only Late Cambrian or just Cambrian age, but cannot be correlated at the zonal level.

It seems extremely difficult to recognize zonal index fossils among the Paraconodontida. Druce & Jones (1971) also had been unable to erect even assemblage zones for this time interval. Further, the total age-ranges of most taxa have probably not yet been established. For this reason we make no attempt at a worldwide conodont-based correlation for this time interval.

A different method for such a purpose may be the application of eustatic changes of the sea level. Miller (1978, 1984, 1988a, b) and Erdtmann & Miller (1981) discussed them as possible reason for a change in the faunal composition. Miller (1984, 1988a, b) utilised a worldwide eustatic model to explain two sequences of shallow water high-energy sediments, associated unconformity surfaces, and trilobite and conodont extinction.

A lowering of sea level can also be observed in the Upper Cambrian series of Norden. The *Acerocare* limestone commonly is developed as a higher-energy sediment, which is very rare in the earlier zones. Compared with the *Peltura* Zone, these deposits are obviously better oxygenated. If this shallowing is connected with a worldwide regression, the *Acerocare* Zone would have to correlate with the *Hirsutodontus hirsutus* subzone of the *Cordylodus proavus* Zone *sensu* Miller (1988a). The suggested rise in sea level in the *Fryxellodontus* Zone or somewhat later accords with the *Dictyonema* Shale in Scandinavia (compare Miller 1984, Fig. 1)

which obviously was deposited in deeper water again. It would also be consistent with a deepening-upward cycle in China (Miller 1988a). However, it should not be ignored that a comparison of conodont associations alone is insufficient to recognize such events.

Factors limiting species distribution

The large fauna presented in this paper has been sampled from sites which are regionally distant from each other but which are characterized by a similar lithology all over the basin. They contain abundant and diversified proto- and paraconodonts associated with already highly developed euconodonts. Differences in the faunal composition compared with other areas may have various reasons:

Age. – The Swedish zones may be incomplete and the missing time intervals may be developed elsewhere or vice versa.

Environment. – Conodonts are considered to have been free-swimming organisms and thus were not strictly confined to certain lithologies. However, conodonts may have been restricted to certain areas within the water column and accordingly would have become deposited as part of thanatocoenoses in quite different areas.

Facies. – In the Swedish Upper Cambrian the prevalence of single species like *Agnostus pisiformis* in zone I and *Orusia lenticularis* in zone III indicates the development of a special facies. Even if it is easily recognizable over wide parts of the basin, it is not known from other areas.

Water, temperature and geographic position. – The temperature is mainly controlled by two factors: water depth and geographic latitude. The latter is quite a speculative field. Scotese *et al.* (1979) pointed out that paleolatitudes cannot be determined before the Late Paleozoic. On their maps, the earlier continents like Laurentia, Siberia and Kazakhstania have been placed in the equatorial position only for lack of evidence. Again, Bond, Nickerson & Kominz (1984) have not been able to produce more than a 'speculative computer reconstruction' based on 'limited paleomagnetic data'.

Miller (1984, 1988a) proposed a warm and a cold water realm for the different distribution of certain conodont taxa. In his interpretation, proto- and paraconodonts are confined to cold-water realms such as Scandinavia, Great Britain, Turkey, Iran, South China and deep-water areas along the margins of North America, India, Kazakhstania and possibly other low-paleolatitude land masses. Miller (1988a) considered the environment, i.e. platform or slope deposits, the most important factor for a similar faunal composition rather than geographic proximity. Müller's study in Iran (1973), however, is based on shallow-water sediments. Both lithology and the great variety of fossils, particularly of the trilobites, suggest warm water conditions rather than the supposed cold water realm.

According to Miller, euconodonts were adapted to warm mid- to low-paleolatitude areas. Chen & Gong (1986) supported this idea by noting the inferred simultaneous occurrence of the *Proconodontus* lineage in North America, Australia and China. On the other hand, findings of *Proconodontus* in the supposed cold-water realm of Sweden together with other euconodonts and an abundant proto- and paraconodont fauna contradict this theory. *Teridontus nakamurai* may serve as another example. This species occurs in the lower *Cordylodus proavus* Zone in cratonal deposits from North America. The same species or a similar one arises much earlier on the Australian continent (Miller 1980). Miller referred to material by Druce & Jones (1971), which was, however, not published by them. The Australian occurrence is a transgression sequence and thus should belong to the shallow-water facies similar to the North American platform. About as early as in Australia and Korea, *Teridontus nakamurai* was recorded from the *Proconodontus* Zone in deeper-water sequences of continental slope deposits from Newfoundland (Fortey, Landing & Skevington 1982), from Kazakhstan (Chugaeva & Apollonov 1982; Apollonov, Chugaeva & Dubinina 1984) and from China (Wang 1985b; Zhao 1986). According to Landing (1983) such deposits lack taxa which are typical for the inner carbonate platform. However, Landing's fossil lists are not convincing in this respect. Many of his recorded taxa are only tentatively determined or new and thus unsuitable for such comparisons.

In our opinion the evidence for a differentiation of water depth and temperature is still insufficient for a general application at present. Different ranges of species in warm-water and cold-water realms may be simply due to lack of evidence. Apart from that, the application of such concepts requires absolutely reliable determinations on the species level. In some cases the gross morphology in conodonts is insufficient for even generic determination without knowledge of the internal structure. Faunal lists of previous papers are hardly based on such examinations; quite often the state of preservation limits precise determinations.

Salinity. – Salinity as a distributional factor has been studied in detail, e.g., for Ostracoda and Foraminifera. Although it has not been discussed in detail in connection with conodont occurrences, this factor should be taken into consideration. Miller (1984) inferred a tolerance of higher salinity for *Clavohamulus, Cordylodus, Eoconodontus, Fryxellodontus, Hirsutodontus, Monocostodus, Semiacontiodus,* and *Utahconus.* At present, comparative studies are lacking which would enable a generalisation or rejection of this concept.

Nutrition and geochemistry. – Faunal diversity may also depend on a sufficient or deficient availability of nutrition and mineral components such as phosphate. Oxygen tolerance and further factors, such as the concentration of heavy metals (Miller 1987, 1988b), may be relevant as well.

Provincial or endemic occurrence. – Provincialism as can be observed in the extant faunal distribution is a regional differentiation between associations that is independent from physical differences. In the Cambrian this phenomenon is particularly distinct for trilobites. For conodonts, An (1981) stated that a 'conodont provincialism is not clear due to primitive state of evolution', contrary to various authors who suggested a division into a North Atlantic and a North American Midcontinent Province (Fåhraeus & Nowlan 1978; Miller 1984, 1988b; Dong 1987; and preliminarily also Bergström 1988). As this division is based on physical factors such as water temperature, depth, salinity etc. it cannot be considered as relevant for the development of provincialism.

Collecting bias. – It is unlikely that small collections, particularly if only from a single or few outcrops, represent the fauna comprehensively. The confidence, above all for long-range correlations, is considerably reduced by such deficiencies.

For an interpretation, all these possibilities have to be considered and weighed against each other. The numerous possible combinations make it rather difficult to recognize the factors which are relevant in each case for the appearance or disappearance of certain forms.

Preservation

Despite the systematic consistency of the material, the state of preservation is quite variable. This results from physical conditions to which the rocks have been exposed in certain areas as well as to the organic content of the samples. Mechanically, the conodonts do not seem to be much affected. Sorting by currents is not obvious, and abrasion due to transport has been observed only on few specimens but is not a predominant feature.

The occurrence of fairly large clusters indicates that the individual sclerites must have been preserved *in situ*, being glued together by phosphatic matrix. An apparent character in this context is the consistently closed position of the apparatus. In the open one, the sclerites would have been out of contact with each other and would have fallen apart postmortally. Survival of the mainly protoconodont clusters in our material also may have been due to coating with phosphatic matter, e.g., in *Phakelodus elongatus* (Pl. 1:7, 8, 12). Alternatively, clusters may be preserved in coproliths (Pl. 41:11, 14, 16). Where the original arrangement is lost, the disjunct sclerites are randomly orientated as in *Trolmenia acies* (Pl. 26:11). In the *orsten*, this type is, however, rather rare.

In general, fine-grained to fairly coarse phosphatic coatings are frequently observed on both the fossils and sedimentary structures.

Histological studies among the different taxa are important for the recognition of major differences of suprageneric rank. The individual preservation potential of single layers seems to be different, too. Genera such as *Phakelodus, Gapparodus,* and *Gumella* may exhibit stripe-like deposits on their outer surface. They may be regarded as relicts of a previously complete outer layer caused either by resorption as phosphate- and/or weight-saving device or by slightly different resistance to wear.

A rare but striking phenomenon which has been observed particularly on paraconodonts is a large space between the lining of the basal opening and the outermost layer (Pl. 17:12). It might be due to dissolution or shrinkage of the former lining, as is also suggested by an undulation of the lamellae. Interspaces originated in this way may possibly have been filled secondarily with phosphate or calcite, the latter being etched out during the preparation process.

In genera such as *Furnishina*, *Nogamiconus*, *Problematoconites*, and *Westergaardodina*, semicircular bulgings on the flanks have been noted (Pls 13:5; 21:11). Apparently they are not set off from the outermost layer. In a few cases they are longitudinally or partly concentrically ruptured. It is unclear whether these structures are pathologic or whether they are mere artifacts, perhaps originated by crystal growth.

According to Clark & Miller (1969) there has been an evolution from the earliest organic-rich representatives to sclerites with a massive apatite fabric. If the early sclerites consisted only of isolated apatitic rods in an organic matrix, those specimens would fall apart due to, e.g., oxidation of the organic substance. The more well-developed conodonts, however, would not be affected. Such differences in the preservation potential influence the faunal record.

An apparent characteristic of the Swedish conodonts is their colour range from translucent light yellow tints up to totally black. At first glance they seem to represent several stages of thermal alteration. But paraconodonts cannot and should not be used for colour alteration index determinations, because their thermal maturation behaviour has not been studied, nor calibrated experimentally. The technique is only applicable to euconodonts.

Our fauna is embedded in finely dark and light coloured *orsten* (cf. also Szaniawski 1971). A possible explanation for the co-occurrence of amber-coloured and black specimens in the same samples may be a secondary staining within the dark layers.

In addition, there are specimens with the surface eroded and altered into a whitish colour. Three alternative explanations are possible: (1) A weathering in surface samples with its greatest effect on material of the isle of Öland. (2) An alteration into translucent-white originating from an oxidation of the organic matter prior to sedimentation. (3) The alteration may also document a certain stage of the etching process. The fresh, unbuffered acid dissolves all phosphatic matter exposed on the surfaces of the crushed rocks. In an intermediate stage, phosphatic fossils may be more or less altered. Only in the final process, with the buffer solution being fully established, do the specimens remain in their original condition.

Clusters, assemblages and apparatuses

Clusters are sclerites preserved in their natural context. In general they represent only part of an apparatus. They are known throughout the whole group from Cambrian to Triassic times. Upper Cambrian clusters reflect mainly unimembrate apparatuses. They occur either on bedding planes in shale facies or in etched residues, fused by phosphate or silica matrix. Contact between the individual sclerites is essential for clusters in residues. This occurs, e.g., in the contracting phase of *Phakelodus* clusters (Szaniawski 1982). As the sclerites are not fused during life, the open, expanded phase of the apparatus has only little preservation potential.

By contrast, assemblages are parts of uni- or multimembrate apparatuses which have been reconstructed mainly from isolated specimens and may include morphologically quite diverse forms which were previously regarded as different genera. Comparative studies with known apparatuses from, e.g., the Devonian and Carboniferous, together with the application of statistics, have been a means for their recognition. According to van Wamel (1974), multielement species can be reconstructed on the basis of similar stratigraphic range, comparable relative frequencies, similar colour and/or mineral composition, as well as similar morphological features. This is, however, still difficult for the Cambrian. For example, Landing, Ludvigsen & von Bitter (1980) and Fortey, Landing & Skevington (1982) considered *F. furnishi* and *F. asymmetrica* as an assemblage. In our material, these two forms appear in different zones – an obvious contradiction to their suggestion.

Cluster research in the Cambrian was initiated by Miller & Rushton (1973) with findings of *Prooneotodus tenuis* (now *Phakelodus tenuis*); Müller & Andres (1976) illustrated a natural assemblage in an Upper Cambrian shale from Sweden. Based on material from eastern New York Landing (1977) reconstructed an apparatus comprising 26 elements organised in two 13-element half-apparatuses. Tipnis & Chatterton (1979) illustrated an almost complete organ as *P. tenuis* from the Northwest Territories, Canada. However, the cross-section of the elements is not consistent with the diagnosis of the species. It is tear-shaped with a keeled posterior side, and the specimen should be referred to *Phakelodus elongatus* n.sp. Szaniawski (1980b) noted clusters of *Prooneotodus gallatini*, a juvenile stage of *Furnishina*, and *Phakelodus tenuis* in the Upper Cambrian of Poland. The paraconodonts comprise only two elements and thus might have been fused accidentally. Szaniawski did not discuss them further in his follow-up papers on the same topic. Because of their general outline, paracondontid elements could not be in contact with each other over the entire length, not even in the contracting phase of the apparatus. Thus adhesion in clusters is extremely difficult. The known clusters are mostly small stages prior to the development of a basal flaring.

Besides the common and widespread *Phakelodus* representatives, the present material has yielded a fairly large variety of unimembrate clusters from the following taxa: *Coelocerodontus bicostatus*, *Furnishina* sp., *Muellerodus subsymmetricus*, *Nogamiconus falcifer*, *Prooneotodus gallatini*, and *Trolmenia acies*. The number of individual sclerites ranges from 2 to at least 26; the latter represents most probably a complete apparatus (Pl. 1:2–4, 13).

As the clusters are only gently stuck together by phosphatic matrix, they easily break apart. This excludes the possibility that individual sclerites are fused with their outer layer 'at least in the basal part' as has been suggested by Tipnis & Chatterton (1979). Fused sclerites with a joint basal opening similar to some of the early shelly fossils do occur, but are considered not to be equivalent to clusters.

Clusters have been assumed to be composed of two crescent-like half-apparatuses facing each other (Landing 1977). But as can be seen on Pl. 1:13, the original shape consists of a more or less circular arrangement. The crescentic appearance is only a matter of preservation due to deformation by compaction and torsion (Pl. 2:15). The arrangement of individual elements within an apparatus seems to be a stable feature. In quartering the circle, the largest conodonts are positioned in the centre of each sector. This documents that certain positions within the apparatus require a definite size of the sclerite which may differ from an adjacent one. While small elements or groups have tentatively been considered as juvenile (Szaniawski 1980b), it is hardly possible to distinguish between mature and immature stages on isolated sclerites. Furthermore, we lack convincing evidence for a successive establishment of new sclerites during growth of the animal.

Assemblages are difficult to recognize in our material. As has been documented for clusters, the sclerites in an apparatus are principally of the same shape and differ mainly in size according to their position in the organ. The cross-sections of the unimembrate *Ph. elongatus* and *Ph. tenuis* (Pl. 1:12, 15)) may serve as examples. By contrast, multimembrate apparatuses can be identified only if their various morphotypes are not intergrading. This is the case in *Cambropustula kinnekullensis*, *Coelocerodontus bicostatus*, *Gumella cuneata*, and *Nogamiconus falcifer*. Accordingly, we refrained from apparatus reconstruction for all other taxa.

Function

A comprehensive account of previous palaeobiological interpretations is given by Repetski & Szaniawski (1981). In the following, only some supplementary discussion is added.

Growth of para- and euconodonts is so distinct that there cannot be a generally accepted functional interpretation. Paraconodonts grew by basal internal accretion, contrary to the outer apposition of euconodonts. However, neither case permits an interpretation as teeth with a chewing function as was originally proposed by Pander (1856).

For the slender, protocondontid elements *sensu* Bengtson (1976), Szaniawski (1982) suggested a tooth function similar or identical to chaetognaths, i.e. grasping, grabbing, or holding. An alternative interpretation considers these forms as filter organs or acting as traps. From a great many clusters, the individual sclerites are known to be arranged in a circle, similar to Recent chaetognath grasping organs. The apparatus was able to open and close repeatedly, which points against a supertooth function as suggested by Landing (1977). It seems to us that the latter interpretation has been prompted by a phosphatic film covering tightly attached elements. Occasionally, three-dimensional, almost complete apparatuses are preserved this way.

However, most of the other paraconodont representatives have not been discovered in coherent apparatuses. Judging from the individual shape of the sclerites, their function must have been different from that of protoconodonts, but we still lack evidence on their use.

An interpretation as a lophophore-supporting organ of *Odontogriphus omalus* from the Burgess Shale (Conway Morris 1976) is rather unlikely, because its small size would be insufficient for the nutrition of such a comparably large animal.

Euconodonts, with their outward directed growth, must have been embedded in tissue during the entire growth period(s). This points to a function as internal sclerites. Bengtson (1976) introduced a hypothesis, differentiating between a functional and resting position. Growth would have taken place periodically within pockets, and the intermittently emerging elements could then act as external sclerites.

A uniform function for all conodonts is not likely, neither within the various orders nor necessarily for multimembrate elements of the same animal. Possible functions refer to teeth, supporting organs, or chemical deposits, with the latter being resorbed in time of need. Elements with a similar shape can be functional homeomorphs. This was already suggested by Smith, Briggs & Aldridge (1987) who interpreted the gross similarity of the euconodont *Panderodus* apparatus for a similar function as *Phakelodus*.

Systematic position

The systematic position of the Conodonta has been discussed since 1856, and more than 50 assignments have been published. In recent years, the discussion has centered around two theories relating conodonts either to chordates or to chaetognaths.

Based on the soft-part morphology, the conodont animal from the Carboniferous of Scotland has been placed into a group of jawless craniates by Aldridge *et al.* (1986). Dzik (1986) also favours a chordate affinity of the conodonts. However, according to the former authors, a specimen from the Silurian of Waukesha, Wisconsin '... would imply a body form more comparable with that of chaetognaths than is the laterally flattened form deduced for the Edinburgh specimen'. The theory of chaetognath affinity stresses the morphological and structural similarities between protoconodont elements and grasping spines of chaetognaths (Szaniawski 1980a, 1982, 1987; Bengtson 1983b).

Since Bengtson differentiated a protoconodont growth type within the paraconodonts, it has been debated whether protoconodonts can be regarded as conodonts at all. Sweet (1985), e.g., stated that 'the protoconodonts are chaetognaths, not conodonts. They may be trying to tell us something about the affinities of the conodonts, and may even suggest that chaetognaths are reasonable living analogues for interpretation of certain conodont features, but

I doubt that they are telling us that conodonts are chaetognaths.'

Presently it is still unclear whether protoconodonts have to be excluded from true conodonts. Bengtson's interpretation of an evolutionary sequence from proto-, through para- to euconodonts seems to be fairly plausible. All sclerites are composed of an organic matrix with an inlay of apatitic crystals. There is a remarkable uniformity in the development of structures, e.g. furrows, denticles etc., within the phylum. A consistent tendency from simple to complicated shapes is evidenced. This is well in accord with a chemical evolution from early sclerites with high organic and little phosphatic content to forms with gradually increased phosphate and decreased organic matter. Transitions from para- to euconodonts have been repeatedly reported (Miller 1976, 1980; Szaniawski & Bengtson 1988). A histological development from proto- to paraconodonts, i.e. from more or less straight extending to outward directed cuffs, seems likely but has not been proved yet. This problem requires extensive thin-sectioning, focusing particularly on this problem. Further, the appearance of proto- and paraconodonts in the geological column is consistent with their respective level of structural and morphological development. Protoconodonts are known from the Precambrian–Cambrian boundary to the lowermost Ordovician. Paraconodonts occur from Middle Cambrian to Middle Ordovician times; euconodonts arise in the basal Upper Cambrian and become extinct in the Upper Triassic. During the Cambrian, the time of stratigraphic overlap, it is a striking phenomenon that all three types commonly occur together.

Systematic palaeontology

The main character for systematic subdivision is the mode of growth. Two different types can be distinguished:

(1) Basal–internal accretion as observed in the Paraconodontida. Bengtson (1976) recognized a structure on *Amphigeisina* and *Gapparodus* which was different to the yet known paraconodonts with their basally cuffed, i.e. upwardly outcropping, lamellae. The new growth type, termed protoconodont, is characterized by more or less straight lamellae which form an acute angle with the surface. The basal margin is extremely thin. It is composed of the innermost lamella only and thus hardly ever preserved on etched specimens. Protoconodonts have been considered as a subgroup of the Paraconodontida as they agree in their overall direction of growth.

Already in this group a development from coniform to compound sclerites can be observed. This was attained in different ways: in *Westergaardodina* it took place by an upraising of the lateral projections. *Serratocambria* developed numerous small denticles on top of the lateral process by differentiation during growth or by resorption.

Paraconodonts were believed to represent or to be homologous with basal bodies of conodontophorids (Druce & Jones 1971; Bengtson 1976; Szaniawski 1980a; Andres 1981; Müller *in* Robison 1981). Following this interpretation, the generally large basal invagination of paraconodonts would not be homologous to the basal cavity of euconodonts, but rather to the lower surface of the basal body. For this reason, we use the non-committal term 'basal opening'. These early forms, however, have a strikingly diversified morphology not found in basal bodies of subsequent euconodont taxa. Furthermore, on some elements of *Furnishina rara*, *Furnishina sinuata*, and *Problematoconites* (Pls 15:12; 17:6; 23:21, 22, 24), the inner side is fairly thick and dotted with irregularly arranged pits, which could be interpreted as further histological differentiation of a basal body.

(2) The conodontophorid mode of growth is characterized by outer apposition with an increasing differentiation of the individual lamellae. The development of white matter, which has not been proved for paraconodonts yet, is an additional character. The presence of the already fairly advanced new genus *Cambropustula* in the lower Upper Cambrian supports the assumption of a polyphyletic origin of the group (see also Miller 1984).

The early conodonts display strikingly diverse internal features; a detailed study of various genera is under way. Particularly for the determination of coniform elements, the position of three features, which do not necessarily coincide, may be systematically relevant: (1) the tip of the basal cavity, (2) the end of the costae, and (3) the point of flexure. Examples are the genera *Furnishina* and *Muellerodus*.

For a taxonomic study, intra- and interspecific variability should also be taken into consideration. In general, variability refers to the individual development of homologous members in different apparatuses of the same species. It is always restricted to certain features and its extent varies from species to species, as referred to in the descriptions.

In addition, our unimembrate clusters are characterized by differences between the specimens according to their different positions. On isolated sclerites, such differences cannot be exactly referred to either the position within an apparatus or to different organs at all. For this reason, such respective elements have to be regarded as intraspecific varieties.

Intraspecific variability, particularly on little-differentiated sclerites, may obscure species boundaries even on well-preserved material. Such cases can be referred only to the genus or, at most, to one or another species (species A or B).

In the present monograph, the taxonomy in the rank of orders suggested in the *Treatise on Invertebrate Paleontology* is being maintained.

Morphological terminology

The morphology of elements described here is based on the defintions given in the *Treatise on Invertebrate Paleontology*, pp. W5–W16. Terms newly introduced or modified are as follows:

Annulation: ribs more or less parallel with the basal margin that are caused by thickening or undulation of the

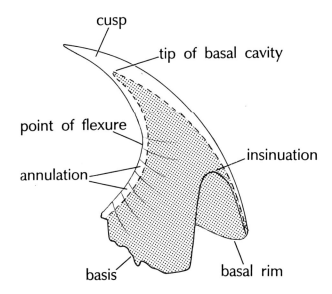

Fig. 5. Explanation of morphological terms for simple cones.

Line of connection: line of incomplete closure by two adjacent faces, leading to a keel-like structure in *Nogamiconus sinensis* (Fig. 12).

Point of flexure: deepest point in the recurvature of the posterior side (Fig. 5).

Serrae: denticle-like differentiation (Fig. 15).

Sheet: thin but solid connection between posterior face and keel in *Westergaardodina* (Fig. 6).

Top end: termination of lateral projections in *Westergaardodina* (Fig. 6).

Tunnel: tube-like structure on the posterior side of a conodont. It may be part of the basal opening as, e.g., on some species of *Westergaardodina* and *Furnishina*. This structure obviously occurs in groups which are systematically distant. Therefore it is believed to be a derived (plesiomorph) character.

Turning point: deepest point between adjacent processes in *Westergaardodina* (Fig. 6).

lamellae (compare *F. primitiva, M. pomeranensis, Ph. simplex, P. gallatini*).

Basal opening: basal invagination of paraconodonts that may be considered homologous to the lower surface of the basal body in euconodonts.

Basis: lower rim of a specimen as preserved.

Bridge: more or less narrow part posteriorly in the mediobasal plane of bicuspidate *Westergaardodina* where anterior and posterior sides are fused (e.g. *W. matsushitai*, Pl. 28:16, 20).

Callosity: emerged, crudely folded structure around the turning points of *Westergaardodina*.

Indentation: arched basal margin with sharp-edged top(s) in *Prosagittodontus* (e.g. Fig. 14).

Insinuation: differentiation of a side by rounded arching of the basal margin (e.g. Fig. 5).

Lamina: thin, flat extension on a costa originated by fusion of both anterior and posterior portions of the sides, e.g., on *Furnishina, Proacodus* and *Coelocerodontus* (Fig. 13).

Order Paraconodontida Müller 1962

Genus *Bengtsonella* n.gen.

Type species. – *Bengtsonella triangularis* n.sp.

Derivation of name. – In honour of Dr. Stefan Bengtson, Uppsala.

Diagnosis. – Coniform, probably paraconodont elements with a narrow, rounded costate anterior side. The posterior face terminates at similarly rounded costae. A distinct recurvature is combined with a slight lateral deflection.

Bengtsonella triangularis n.sp.
Pl. 6:1–7; Fig. 7A, B

Derivation of name. – Latin *triangularis*, after the triangular basal opening.

Holotype. – UB 1006 (Pl. 6:3–5).

Type locality. – Stenstorp–Dala.

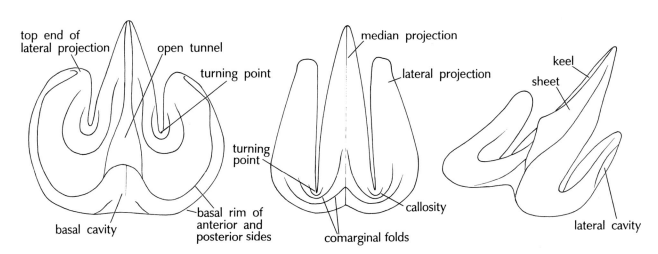

Fig. 6. Explanation of morphological terms for *Westergaardodina.*

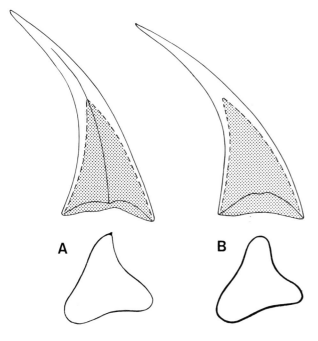

Fig. 7A, B. Bengtsonella triangularis n.gen., n.sp. Outline with basal opening and cross-section; all ×170.

Type horizon. – *Peltura scarabaeoides* Zone (Vc).

Material. – 50 specimens.

Occurrence. – Zone V: Degerhamn, Grönhögen, Gum, S. Möckleby, Skår, Stenstorp–Dala, St. Stolan.

Diagnosis. – As for the genus.

Description. – Slender asymmetrical simple cones which are distinctly recurved and somewhat laterally deflected. The anterior side is rounded costate. The posterior one is flat to slightly concave and bounded by widely rounded posterolateral costae which may also be sharpened (Pl. 6:2; Fig. 7A). The flanks are asymmetrically concave. The large basal opening occupies about half of the entire length. The basal rim is smooth and evenly developed. The cross-section changes from circular at the apex into subtriangular at the aperture. The outer surface may be annulated on the basal portion. The inner one appears smooth.

Size range. – 410–670 μm.

Comparison. – The basal outline is just reverse to the genus *Furnishina* with its flat anterior side. Furthermore, the costae are broadly rounded here in contrast to the generally sharp structures of *Furnishina*, except for *F. primitiva* and *F. sinuata*.

Genus *Furnishina* Müller 1959

Type species. – *Furnishina furnishi* Müller 1959

Furnishina is characterized by distinct, rounded to lamina-like anterolateral costae. A posterior keel may be present. The anterior side always extends beyond the posterior one. The depth of the basal opening exhibits considerable variation. The inner surface is marked by the paraconodont growth lamellae, the outer surface is either smooth or faintly annulated. Left and right forms occur in almost equal frequency.

In his original description Müller erected five species to encompass the various forms of the genus. Our new collections are many times as large as the previous material and permit a more refined species concept.

Proscandodus Müller & Nogami 1971 originally comprised three species: *oelandicus, rarus* and *tortilis* (type species). Szaniawski (1971) referred *P. oelandicus* to *Muellerodus*, an assignment supported herein. In the present paper, both *P. rarus* and *tortilis* are transferred to *Furnishina*. Accordingly the genus *Proscandodus* is now regarded as a junior synonym of *Furnishina*.

Furnishina alata Szaniawski 1971
Pl. 7:5, 11, 14, 16, 17; Fig. 8H, L

Synonymy. – □1971 *Furnishina alata* n.sp. – Szaniawski, pp. 406–407, Pls. 1:3, 4; 3:3–5.

Material. – 190 specimens.

Occurrence. – Zone I: Backeborg, Gössäter, Gudhem, Gum, Haggården–Marieberg, Kakeled, Kleva, Sätra.

Description. – Small erect sclerites characterized by wing-like extended flanks. The anterior side is flat to slightly convex. Its flared basal section is distinctly set off from the cusp. Anterolateral costae that are lamina-like basally, narrow rather quickly and disappear on the cusp. The posterior side is marked by a keel which either extends to the tip or terminates within the upper fifth of the cone. From the lateral laminae the posterior side rises to form a triangular basal opening. The basal rim is smooth and even. The outer surface appears smooth, the inner one shows densely set growth lines.

Size range. – 540–600 μm.

Comparison. – *Furnishina alata* differs from *F. bicarinata* in the development of the anterior side which represents the greatest width and in the lamina-like lateral extensions.

Furnishina cf. *alata* Szaniawski 1971
Pl. 7:18, 20

Synonymy. – □1987 *Furnishina* aff. *furnishi* Müller – An, Pl. 3:8.

Material. – 2 specimens.

Occurrence. – Zone I: Sätra.

Description. – These paired specimens agree in their general outline with *F. alata*. Their anterior side is widely but asymmetrically flared, the posterior one projects in its lower third. The flanks are lowered almost down to the anterior basal margin and are laterally attached to it. Different from *F. alata* is the position of the posterior keel, which is not developed on the base but extends all along the cusp up to the tip.

Size range. – 350–400 μm.

Furnishina asymmetrica Müller 1959
Pl. 10:1–19; Fig. 8F

Synonymy. – ☐1959 *Furnishina asymmetrica* n.sp. – Müller, pp. 451–452, Pl. 11:16, 19. ☐1971 *Furnishina asymmetrica* Müller – Müller, p. 8, 17, Pl. 1:13; Fig. 1b. ☐1986 *Furnishina quadrata* Müller – Chen & Gong, p. 147, Pl. 17:17; Fig. 52/2.

Material. – 560 specimens.

Occurrence. – Zone I: Backeborg, Gudhem, Gum, Klippan, Sätra; Zone II: Haggården–Marieberg, Toreborg; Zone III: Karlsfors, Mossebo, Nygård, St. Stolan; Zone IV: Grön-högen, Nygård; Zone Vb: Stenstorp–Dala; Zone Vc: Grön-högen

Description. – Simple sclerites which are straight or pro-clined to a variable extent; lateral deflection has been observed, too. The anterior side is characterized by a flat-tened median costa which widens towards the basal rim and makes the whole side appear undulated. The latter is bounded by blade-like anterolateral costae which are trace-able up to the apex. In a few cases they may appear some-what serrated (Pl. 10:10, 13, 19). The posterior side is convex and carries a strong median costa which extends to the tip. Basally it becomes widely rounded and thus indis-tinct in the cross-section. The basal rim is smooth and may be scalloped. The cross-section changes from circular to the typical asymmetrical opening. It varies between irregu-lar quadrangular and subtriangular, sometimes even with a distinct lateral extension. Within the basal opening, growth lines often are exposed. By contrast, the outer surface is smooth. Left and right forms are distinguished by the shape of the basal opening.

Size range. – 460–1000 μm.

Furnishina bicarinata Müller 1959
Pl. 8:1–6, 9; Fig. 8E

Synonymy. – ☐1959 *Furnishina bicarinata* n.sp. – Müller, p. 452, Pl. 12:3. ☐1986 *Furnishina bicarinata* Müller – Chen & Gong, p. 144, Pl. 17:15.

Material. – 80 specimens.

Occurrence. – Zone I: Backeborg, Gum, Sätra; Zone II: Hag-gården–Marieberg, Österplana, St. Stolan, Toreborg; Zone III: Grönhögen, Karlsfors, Mossebo, Nygård, St. Stolan; Zone IV: Nygård; Zone Va: Dwasiden–Hülsenkrug, Mark Brandenburg; Zone Vb: S. Möckleby–Degerhamn, Stenstorp–Dala; Zone Vc: Degerhamn, Grönhögen, S. Möckleby, Stenstorp–Dala; Zone V undiff.: Ödegården, S. Möckleby, Stenstorp–Dala

Description. – Slender sclerites, straight to gently recurved. The anterior side is flattened and delimited by sharp an-terolateral costae which proximally disappear on the cusp. The projecting posterior side carries a keel that fades away on the upper part of the cusp. The flanks of the base are characterized by secondary carinae which represent the greatest width of the elements. The convexity, however, differs considerably between the individual specimens and

is independent of the growth stage. Its development even varies on the same element, leading to an asymmetrical outline. The basal rim is smooth and even. The cross-sec-tion is circular at the initial part and passes into the typical bicarinate outline at the basis. The outer surface is smooth or densely annulated, the inner one shows paraconodont growth lines. Part of the basal organ seems to be preserved on Pl. 8:4.

Size range. – 790–1500 μm.

Comparison. – See *F. alata* and *F. vasmerae*

Furnishina curvata n.sp.
Pl. 13:15, 18, 20, 22–25; Fig. 8I

Synonymy. – ☐?1976 *Furnishina* sp. – Abaimova & Ergaliev, Pl. 14:5.

Derivation of name. – Latin *curvatus*, after the strong curva-ture.

Holotype. – UB 1111 (Pl. 13:25).

Type locality. – Stenstorp–Dala.

Type horizon. – *Peltura scarabaeoides* Zone (Vc).

Material. – 220 specimens.

Occurrence. – Zone III: Nygård, St. Stolan; Zone IV: Nygård; Zone Va: Ödegården; Zone Vb: Ödegården, Stenstorp–Dala; Zone Vc: Degerhamn, Smedsgården–Stutagården, S. Möckleby, Stenstorp–Dala, Trolmen; Zone V undiff.: Öde-gården, Stenstorp–Dala

Diagnosis. – A *Furnishina* which is distinctly recurved so that the apex usually extends beyond the posterior margin. The cusp is longer than in the other representatives of the genus.

Description. – Slender simple cones, strongly recurved over the entire length. Lateral deflection and a slight torsion are common features (Pl. 13:18). The anterior side flattens towards the basis; distinct anterolateral costae are traceable along the base. In lateral view the lower part of the poste-rior side projects nearly rectangularly from the cusp. A keel is developed only along the base. At about its middle it may be marked by a bump (Pl. 13:19) which is similar to a rudimentary denticle. The flanks are subsymmetrical to asymmetrical and appear either flat or slightly bulged. The basal opening is comparatively short and does not exceed the point of flexure. The basal rim is smooth. The circular apical cross-section changes into subtriangular at the basis.

Size range. – 280–570 μm.

Furnishina furnishi Müller 1959
Pl. 13:1–7, 11, 12; Fig. 8A

Synonymy. – ☐1959 *Furnishina furnishi* n.sp. – Müller, p. 452, Pl. 11:5, 6, 9, 11–13, 15; Fig. 6D. ☐1971 *Furnishina as-ymmetrica* Müller – Müller, p. 8, Pl. 1:9, 12, 14–16. ☐1971 *Furnishina furnishi* Müller – Müller & Nogami, p. 14, Pl. 1:5. ☐1972a *Furnishina furnishi* Müller – Müller & Nogami, Fig.

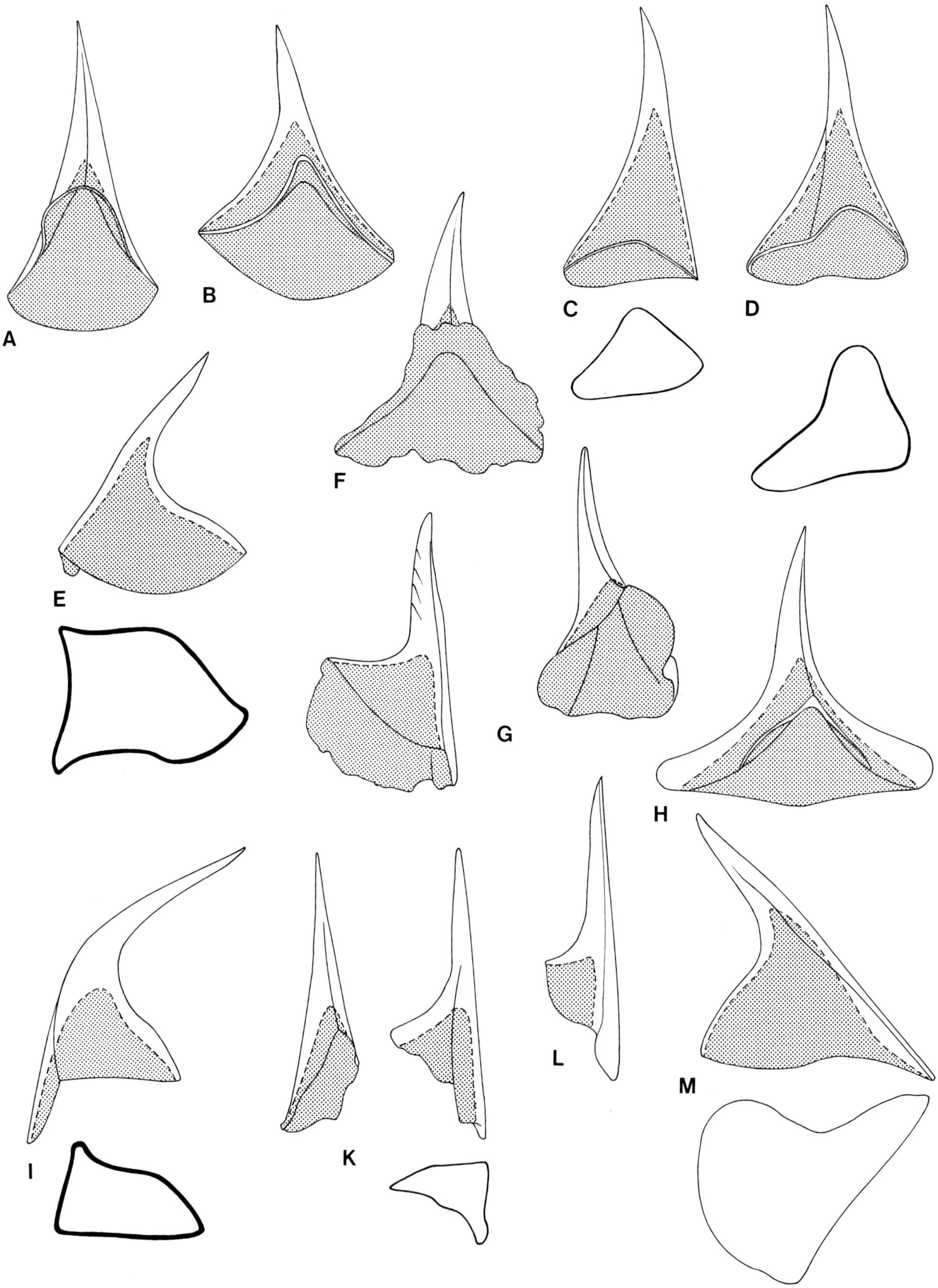

Fig. 8. Outlines of various species of *Furnishina* showing shapes of basal openings and, for some, basal cross-sections. □A. *F. furnishi* Müller 1959; ×100. □B. *F. kranzae* n.sp.; ×55 . □C. *F. primitiva* Müller 1959; ×70. □D. *F. primitiva* Müller 1959; ×55. □E. *F. bicarinata* Müller 1959; ×35. □F. *F. asymmetrica* Müller 1959; ×50. □G. *F. vasmerae* n.sp., ×65. □H. *F. alata* Szaniawski 1971; ×80. □I. *F. curvata* n.sp.; ×115. □K. *F. gossmannae* n.sp.; ×115. □L. *F. alata* Szaniawski 1971; ×80. □M. *F. polonica* Szaniawski 1971; ×100.

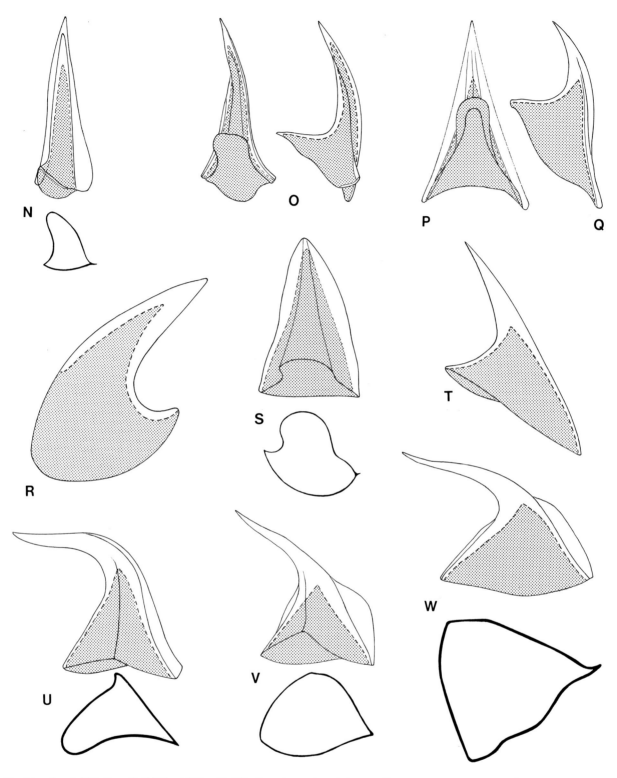

Fig. 8 (continued). □N. *F. tortilis* (Müller 1959); ×85. □O. *F. tortilis* (Müller 1959); ×40. □P, Q. *F. kleithria* n.sp.; ×120. □R. *F. sinuata* n.sp.; ×70. □S. *F. gladiata* n.sp.; ×55. □T. *F. ovata* n.sp.; ×50. □U. *F. rara* (Müller 1959); ×75. □V. *F. mira* n.sp.; ×150. □W. *F. mira* n.sp.; ×160.

1A. □1973 *Furnishina asymmetrica* Müller – Müller, p. 39, Pl. 1:6, 8. □?1976 *Furnishina furnishi* Müller – Abaimova & Ergaliev, p. 391, Pl. 14:1, 2. □?1976 *Furnishina furnishi* Müller – Miller, R.H. & Paden, p. 595, Pl. 1:8–12. □1979 *Furnishina furnishi* Müller – Bednarczyk, p. 427, Pls. 1:1–3, 5–9, 12; 3:14, 20, 21. □1981 *Furnishina furnishi* Müller – Miller, *in* Robison, p. W112, Fig. 64.7a, b. □1985b *Furnishina furnishi* Müller – Wang, p. 90, Pl. 25:7, 8; Fig. 14/11.

□1986 *Furnishina furnishi* Müller – Chen, Zhang & Yu, p. 367, Pl. 1:17. □?1986 *Furnishina furnishi* Müller – Chen & Gong, pp. 145–146, Pl. 52:7. □1988 *Furnishina furnishi* Müller – Lee, B.S. & Lee, H.Y., Pl. 1:16, 17.

Material. – 300 specimens.

Occurrence. – Zone III: Karlsfors, Nygård, St. Stolan; Zone IV: Nygård; Zone Va: Ödegården; Zone Vb: Ödegården,

Stenstorp–Dala; Zone Vc: Degerhamn, Grönhögen, Smedsgården–Stutagården, S. Möckleby–Degerhamn, Stenstorp–Dala; Zone V undiff.: Brattefors, Ödegården, Stenstorp–Dala, Trolmen

Previously, *Furnishina furnishi* was used as pool for many varieties which could not be referred properly to another species. Accordingly its range comprises all zones of the Upper Cambrian to the lowermost Ordovician. Restudy of that particular form has led to a more restricted concept, with the appearance of this species in zone III.

Description. – Slender, erect to proclined simple cones. Often the apex appears shifted rather than deflected to either side (Pl. 13:3, 12). A slight torsion is frequently observed. The anterior side is flattened and terminates anterolaterally at sharp costae which are traceable over approximately three fourths of the entire length. The posterior side is domed, with a median keel of variable extent. The subsymmetrical flanks are flat to convex, forming a subtriangular basal opening. In a few cases the latter appears even bell-shaped. The basal rim is smooth. The cross-section is rounded apically and passes into triangular at the basis. The outer surface may show a faint annulation, the inner one has the growth lines exposed. Intraspecific variations comprise mainly recurvature, the width:height ratio of the basis and the shape of the basal cross-section.

Size range. – 470–870 μm.

Furnishina gladiata n.sp.
Pl. 16:1–13, 15; Fig. 8S

Synonymy. – ☐1959 *Scandodus* n.sp. b – Müller, p. 464, Pl. 12:5.

Derivation of name. – Latin *gladius*, sword.

Holotype. – UB 1143 (Pl. 16:11, 12, 15).

Type locality. – Gum.

Type horizon. – *Agnostus pisiformis* Zone.

Material. – 30 specimens.

Occurrence. – Zone I: Backeborg, Gum; Zone II: Degerhamn, Haggården–Marieberg.

Diagnosis. – Extremely thin-walled specimens with a very large basal opening. The posterior carina opens basally to a broad tunnel-like structure. The anterolateral costae appear as large laminae.

Description. – Generally fairly large elements. Despite their thin wall structure they appear rather stout. They are gently and evenly recurved. Lateral deflections of various degree have been observed, too. The anterior side is broadly rounded, the posterior one is characterized by a median carina in the apical part which widens to a broad tunnel with a flattened basal portion. The concave posterolateral flanks terminate in lamina-like costae that meet at the apex. The basal opening is extremely large, as can be observed on the translucent yellow to light amber specimens. It extends up to the initial part, leaving an exceedingly small cusp. The basal rim is smooth. The basal cross-section is

somewhat similar to *Muellerodus pomeranensis* except that both posterolateral parts are concave. Due to torsion of the base, left and right forms are easy to distinguish.

Size range. – 420–900 μm.

Furnishina sp. aff. *gladiata*
Pl. 16:14

Material. – 4 specimens.

Occurrence. – Zone I: St. Stolan.

Description. – These specimens have a tunnel-like median structure similar to *F. gladiata*. They differ, however, in the large, almost rectangular apical angle; therefore they are only tentatively referred to the species.

Size. – About 1100 μm.

Furnishina gossmannae n.sp.
Pl. 13:8–10, 13, 14, 16, 17, 21; Fig. 8K

Synonymy. – ☐1973 *Furnishina asymmetrica* Müller – Müller, p. 39, Pl. 1:8. ☐1986 *Furnishina furnishi* Müller – Jiang et al., pp. 46–47, Pl. 2:12. ☐1987 *Furnishina furnishi* Müller – An, p. 106, Pl. 3:12.

Derivation of name. – In honour of Mrs. Annemarie Gossmann, Bonn.

Holotype. – Pl. 13:10, 16, 17.

Type locality. – Ödegården.

Type horizon. – *Peltura scarabaeoides* Zone (Vc).

Material. – 180 specimens.

Occurrence. – Zone III: Mossebo, Nygård, St. Stolan; Zone IV: Nygård; Zone Va: Ödegården; Zone Vb: Ödegården, Stenstorp–Dala; Zone Vc: Degerhamn, Ödegården, S. Möckleby–Degerhamn, Stenstorp–Dala ; Zone V undiff.: Ödegården, S. Möckleby–Degerhamn, Stenstorp–Dala, St. Stolan, Trolmen.

Diagnosis. – Slender, erect to proclined elements. A posterolateral secondary carina on one of the flanks makes the posterior base appear obliquely rectangular.

Description. – Simple erect to gently recurved sclerites which may have their tip slightly turned up. The anterior side is flat to slightly convex and is flanked by sharp costae alongside the base. The posterior side is smoothly projecting and marked by a distinct keel which disappears on the cusp. The flanks are differently developed, which becomes most obvious on larger specimens: one side is concave, the other one carries a secondary carina close to the posterior side. The posterior side is rather straight between keel and carina. The basal rim is smooth. The cross-section is circular at the apex and differentiates increasingly towards the basal margin. The outer surface is smooth, the inner one shows dense growth lamellae (Pl. 13:16).

Size range. – 480–960 μm.

Comparison. – Although this form has a crudely quadrangular base, it differs from *F. quadrata* in having a posterior keel and in the generally asymmetrical outline. It is different to *F. asymmetrica* which has lamina-like anterolateral costae and a broad basal cross-section instead of quite a high and narrow one (Pl. 10:15).

Furnishina kleithria n.sp.
Pl. 15:1–5, 7–10, 16; Fig. 8P, Q

Synonymy. – □1983 *Furnishina triangulata* Xiang & Zhang – An *et al.*, pp. 101–102, Pl. 2:3.

Derivation of name. – Greek *kleithria*, keyhole, after the typical outline of the basal opening. The name is a noun.

Holotype. – UB 1128 (Pl. 15:3, 8).

Type locality. – Gum.

Type horizon. – *Agnostus pisiformis* Zone.

Material. – 150 specimens.

Occurrence. – Zone I: Backeborg, Gössäter, Gum, Haggården–Marieberg, Kakeled, St. Stolan, Trolmen; Zone Vb: Stenåsen ; Zone V undiff.: Stenstorp–Dala.

Diagnosis. – A *Furnishina* with extraordinary keyhole-shaped basal opening. The concave, converging flanks are topped by a somewhat broader, open tunnel.

Description. – Almost symmetrical, broadly recurved, coniform elements. The apex is smoothly rounded. The anterior side, which represents the maximum width of the cone, is rather flat, the posterior one is rounded costate and basally arched up. The anterolateral costae are developed as laminae and differ in width. The characteristic basal opening appears roughly triangular. The concave flanks converge to a short neck, which widens to a tunnel. The latter is clearly distinguishable up to the insertion of the lateral laminae. The basal opening is fairly deep and extends over more than two thirds of the total length (Fig. 8P, Q). The basal rim has not been fully preserved on any specimen.

Size range. – 330–510 μm.

Comparison. – There seems to be quite a close affinity between *F. sinuata* and *F. kleithria.* The latter lacks, however, the typical anterior and posterior indentations and is much more delicate than the thick-walled *F. sinuata.*

Furnishina kranzae n.sp.
Pl. 12:1, 2, 6, 8, 12–14, 18; Fig. 8B

Synonymy. – □1959 *Furnishina furnishi* n.sp. – Müller, pp. 452–453, Pl. 11:8a, b, 14; Fig. 6E. □1982 *Furnishina furnishi* Müller – An, p. 132, Pl. 1:15. □1982 *Furnishina asymmetrica* Müller – An, pp. 131–132, Pl. 2:10, 11. □1983 *Furnishina asymmetrica* Müller – An *et al.*, pp. 98–99, Pl. 2:13. □1986 *Furnishina furnishi* Müller – Chen & Gong, pp. 145–146, Pl. 17:4, 13. □1988 *Furnishina furnishi* Müller – Lee, B.S. & Lee, H.Y., Pl. 1:18.

Derivation of name. – In honour of Mrs. Dorothea Kranz, Bonn.

Holotype. – UB 1084 (Pl. 12:8, 13).

Type locality. – Österplana.

Type horizon. – *Olenus*-Zone.

Material. – 50 specimens.

Occurrence. – Zone II: Gössäter, Haggården–Marieberg, Österplana, St. Stolan, Stubbegården, Toreborg.

Diagnosis. – A *Furnishina* with laterally deflected cusp and distinct anterolateral costae; a posterior one is not developed. The subsymmetrical basal opening is triangular.

Description. – Proclined sclerites having their cusp more or less laterally deflected. The pointed apex is gently turned up. The anterior side is flat to very slightly convex, the narrow posterior one which lacks a keel is bounded by differently developed flanks. They are either variably concave, or convex and concave, respectively. The basal opening extends about half of the entire length (Fig. 8B). The basal rim appears usually smooth except for a single specimen with a scalloped margin. The cross-section changes from circular at the apex into subtriangular at the base. Outer and inner surfaces are smooth. Variation is observed mainly in the degree of deflection, the basal height and the general development of the flanks.

Size range. – 770–1100 μm.

Comparison. – This form differs from *F. asymmetrica* in the development of the anterolateral costae, which are not lamina-like, the non-costate anterior face, the absence of a posterior keel and the lateral deflection of the cusp. Although close to *F. polonica*, it is distinguished from the latter by its lateral deflection, the lack of a lateral rib and the more symmetrical basal opening. It differs from *F. furnishi* in the lack of a posterior keel.

Furnishina longibasis Bednarczyk 1979
Pl. 11:10, 12

Synonymy. – □1979 *Furnishina longibasis* n.sp. – Bednarczyk, p. 427, Pl. 1:1, 4.

Material. – 2 specimens.

Occurrence. – Zone II: Haggården–Marieberg.

Description. – Large, gently proclined sclerites with much extended posterior sides so that the forms may be as high as they are long. The anterior face is flattened and bound by sharp costae which fade away on the cusp. The posterior side lacks a keel. The broadly rounded flanks are slightly concave in the centre, widening to both posterior and anterior sides. The basal rim appears smooth. The cross-section changes from circular to a considerably elongate outline.

Size. – About 900 μm.

Furnishina mira n.sp.

Pl. 17:9, 11, 14–20; Fig. 8V, W

Derivation of name. – Latin *mirus*, marvellous.

Holotype. – UB 1161 (Pl. 17:20).

Type locality. – Gum.

Type horizon. – *Agnostus pisiformis* Zone.

Material. – 100 specimens.

Occurrence. – Zone I: Backeborg, Gum, Klippan; Zone II: Haggården–Marieberg; Zone Va: Ödegården; Zone Vb: Ödegården?; Zone Vc: Stenstorp–Dala; Zone V undiff.: Stenstorp–Dala.

Diagnosis. – Tiny asymmetrical elements with their spine-like cusp deflected towards one of the flanks. The according anterolateral costa opposite this deflection is developed as large lamina.

Description. – Small sclerites which are characterized by a spine-like, strongly recurved cusp. The latter may also be slightly uplifted apically. Further, the cusp is not only shifted out of the midplane, but also slightly twisted towards one of the flanks. The anterior side is asymmetrical in its basal portion. Dominating features are the anterolateral costae which extend up to half length of the sclerite. One of them is developed as a large lamina and continues the line of the cusp down to the basal rim (Pl. 17:20). The other one is a sharp costa. This fabric results in a striking asymmetry. The posterior side is convex with a considerably projecting base. It is marked by a median keel which disappears together with the anterolateral costa. The flanks are gently convex to almost flattened. The basal opening is fairly large and terminates shortly below the level of insertion of the costae. The basal rim is smooth. The cross-section changes from circular at the apex into rounded triangular at the basis. The outer surface shows a faint annulation. Growth lines are distinctly exposed within the basal opening. Left and right forms have been observed.

Size range. – 330–410 µm.

Furnishina ovata n.sp.

Pl. 8:7, 8, 10–22; Fig. 8T

Derivation of name. – Latin *ovata*, after the oval basal outline.

Holotype. – UB 1042 (Pl. 8:16, 19).

Type locality. – Stenstorp–Dala.

Type horizon. – *Peltura scarabaeoides* Zone (Vc).

Material. – 180 specimens.

Occurrence. – Zone Va: Ödegården; Zone Vb: Ödegården, S. Möckleby, Stenstorp–Dala; Zone Vc: Grönhögen, Haggården–Marieberg, Milltorp, Skår, Smedsgården–Stutagården, S. Möckleby–Degerhamn, Stenstorp–Dala, Tomten; Zone V undiff.: Mörbylilla–Albrunna, Ödegården, Smedsgården–Stutagården, Stenåsen, Trolmen.

Diagnosis. – A *Furnishina* with coniform cusp and a large oval basal opening. The anterolateral costae are restricted to the base.

Description. – Simple cones with a gently recurved pointed apex which may also be little deflected. The anterior side is flattened to slightly convex and bounded by the typical anterolateral costae. In other species they ususally represent the greatest width; here they are exceeded by the flanks; this is best observed on large specimens. The costae are restricted to the basal section and are not as pronounced as in related forms. Accordingly they are not exhibited in the basal cross-section (Pl. 8:22). The convex posterior side is non-costate and projects almost rectangularly. The base occupies approximately half of the entire length; the basal rim is smooth. The cross-section of the cusp is circular and changes into quite a variable oval base. Left and right forms occur in the same samples and may be marked by a flattened flank.

Size range. – 320–600 µm. The most frequent stages are 500–600 µm in length.

Comparison. – see *F. quadrata*.

Furnishina polonica Szaniawski 1971

Pl. 11:1–9, 11; Fig. 8M

Synonymy. – □ 1966 *Furnishina asymmetrica* Müller – Nogami, p. 354, Pl. 9:1, 2. □ 1971 *Furnishina polonica* n.sp. – Szaniawski, pp. 405–406, Pls. 1:1; 2:1, 2; 3:1, 2; 5:3; Fig. 1b.

Material. – 450 specimens.

Occurrence. – Zone I: Backeborg, Gudhem, Gum, Haggården–Marieberg, Kleva, Nästegården, Sätra; Zone II: Berlin, Haggården–Marieberg, Klippan, Österplana, Stubbegården, Toreborg; Zone III: Karlsfors, Nygård, St. Stolan; Zone IV: Nygård.

Description. – Straight sclerites with a flattened to plain anterior side. It is flanked by distinct costae which terminate on the cusp. In the basal part they may be developed as laminae. The edges sometimes carry perpendicular ribs and thus appear somewhat serrated (Pl. 11:3). The posterior side is convex and opens to a wide, asymmetrical basis which somewhat resembles that of *F. asymmetrica*. But contrary to the latter, a distinct posterior costa has not been developed. One of the expanded flanks may bear an oblique rib. The coniform apex is pointed and quite abruptly set off from the flared base. The latter is fairly large and occupies more than half of the entire length. The basal rim is smooth. The cross-section changes from circular at the apex into asymmetrical at the basis. Paraconodont growth lines are exposed particularly in the basal opening.

Size range. – 450–720 µm.

Comparison. – *Furnishina polonica* closely resembles *F. asymmetrica* in its basal cross-section; it differs, however, in that the cusp is distinctly set off from the flared base and further the absence of both a posterior keel and an anterior costa. A distinguishing feature might be the oblique rib on the convex flank. This character is, however, not clearly developed on all specimens.

Furnishina primitiva Müller 1959
Pl. 12:3, 4, 7, 9, 10, 15–17, 19; Fig. 8C, D

Synonymy. – □1959 *Furnishina primitiva* n.sp. – Müller, p. 453, Pl. 11:1?, 2–4. □1971 *Furnishina primitiva* Müller – Müller, Pl. 1:11. □1976 *Furnishina primitiva* Müller – Abaimova & Ergaliev, Pl. 14:8. □1985b *Furnishina primitiva* Müller – Wang, p. 92, Pl. 21:15–17. □1986 *Furnishina primitiva* Müller – Chen & Gong, p. 146, Pls. 17:7, 11, 18; 18:11; Fig. 51. □1987 *Furnishina primitiva* Müller – An, p. 106, Pl. 3:10. □1988 *Furnishina primitiva* Müller – Lee, B.S. & Lee, H.Y., Pl. 1:14, 15.

Material. – 1000 specimens.

Occurrence. – Zone III: St. Stolan; Zone Va: Mark Brandenburg; Zone Vb: Ödegården, Stenstorp–Dala; Zone Vc: Degerhamn, Grönhögen, Gum, Skår, Smedsgården–Stutagården, S. Möckleby–Degerhamn, Stenstorp–Dala; Zone V undiff.: Degerhamn, Ekeberget, Ödegården, Skår, S. Möckleby–Degerhamn, Stenstorp–Dala, Trolmen.

Description. – Simple cones that are gently recurved, particularly in their apical part. The cusp is usually deflected to either side. The anterior face is flat and delimited by broadly rounded costae which pass smoothly into the circular cusp at about half length. The posterior side projects only slightly and is characterized by a broadly rounded costa which may tend to a tunnel-structure (Pl. 12:9). The flanks are asymmetrical and appear either flat or concave. The basal rim is smooth and even. The cross-section is circular along the cusp and differentiates into rounded triangular where the costae are set off from the cone. Most of the sclerites have their outer surface distinctly annulated. On the inner surface, traces of paraconodont growth lines are preserved.

Size range. – 390–500 μm.

Comparison. – *Furnishina primitiva* and *F. sinuata* are the only representatives of the genus which lack sharp anterolateral costae. There is a similarity between *Furnishina primitiva* and *Prosagittodontus minimus* in the offset between cusp and base, but *P. minimus* has distinctly developed, sharp lateral costae. Sometimes the cross-section of *F. primitiva* is close to smaller stages of *P. dahlmani*, but the latter is separated into three lobes and additionally has an extremely large basal opening.

Remarks. – Druce & Jones (1971) included the species in *Coelocerodontus*, whereas Fåhraeus & Nowlan (1978) preferred a relation to *Proconodontus*. Both genera, however, belong to euconodonts.

Furnishina quadrata Müller 1959
Pl. 9:1–13

Synonymy. – □1959 *Furnishina quadrata* n.sp. – Müller, pp. 453–454, Pl. 12:2, 4, 9; Fig. 6C. □1966 *Furnishina quadrata* Müller – Nogami, p. 355, Pl. 9:3, 4. □1976 *Furnishina*

quadrata Müller – Abaimova & Ergaliev, p. 392, Pl. 14:3, 4, 6, 7. □?1978 *Furnishina quadrata* Müller – Landing, Taylor & Erdtmann, Fig. 2E. □1979 *Furnishina quadrata* Müller – Bednarczyk, p. 428, Pl. 1:15, 16. □1981 *Furnishina furnishi* Müller – Miller, R.H. *et al.*, Fig. 4E.

Material. – 160 specimens.

Occurrence. – Zone II: Ekebacka, Gössäter, Haggården–Marieberg, Ledsgården, Österplana, Skår, St. Stolan, Stubbegården, Toreborg; Zone III: Nygård, St. Stolan; Zone Va: Ödegården; Zone Vb: Stenstorp–Dala; Zone Vc: Degerhamn, Grönhögen, S. Möckleby–Degerhamn, Stenstorp–Dala, Stubbegården, St. Stolan, Trolmen; Zone V undiff.: Ödegården, Stenstorp–Dala, St. Stolan, Trolmen.

Description. – Distinctly recurved sclerites. The pointed apex may also be slightly uplifted. Some specimens are torted. The anterior side is concave, particularly in the lower portion, and is bounded by well-developed costae which are restricted to the base. In the lower third, the gradually diverging edges of the cone suddenly widen out in a blunt angle. The basally expanded posterior side is convex and widens to a quadrangular opening. A costa or keel is not developed. The flanks sometimes appear undulated as does the basal rim. The cross-section of the cusp is rounded and changes over to quadrangular in the upper part of the base and into the flared shape of the basal margin. In general, there is some variation in the length:width ratio of the basal opening, the degree of asymmetry or the development of processes as well as in the presence of an undulated basal section.

Size range. – 310–1100 μm.

Comparison. – This form is similar to *F. ovata* in its non-costate cusp and the large flared base. A major difference, however, is the concave anterior side. Further, the anterolateral face may be extended a little to either side, but it never appears as a lamina, in contrast to *F. alata*.

Furnishina rara (Müller 1959)
Pl. 17:1–8, 10, 12, 13; Fig. 8U

Synonymy. – □1959 *Scandodus rarus* n.sp. – Müller, pp. 463–464, Pl. 12:12.

Material. – 760 specimens.

Occurrence. – Zone I: Backeborg, Gössäter, Gum, Klippan, St. Stolan; Zone II: Degerhamn, Gössäter, Haggården–Marieberg, Klippan, Österplana, St. Stolan, Stubbegården, Toreborg; Zone III: Grönhögen, Karlsfors, Nygård, St. Stolan; Zone IV: Nygård ; Zone Va?: Ödegården; Zone Vb: Stenåsen, Stenstorp–Dala; Zone V undiff.: Mörbylilla–Albrunna, Ödegården, Stenstorp–Dala, Stubbegården.

Description. – Coniform sclerites that are strongly recurved, particularly in the cusp. The latter is not only twisted but also deflected and tends to be uplifted. The anterior side is flattened and bounded anterolaterally by laminae of different width. The laminae disappear on the cusp. Similar to other species they may bear perpendicular ribs. A posterior keel is not developed. The tongue-like posterior projection is shifted out of the midplane towards the same side as is the deflected cusp (e.g., Pl.17:10). Accordingly, the flanks which decline either flat or concave towards the costae, are asymmetrical. The large base terminates shortly beyond the abrupt recurvature. The basal rim is smooth. The cross-section passes from circular across the cusp into subtriangular at the basis. The outer surface is usually smooth. On one specimen the basal organ is preserved as thickened inner unit which is dotted with irregular pits (Pl. 17:6).

Size range. – 470–1200 μm.

Comparison. – This form is closest to *F. obliqua* in its twisted anterior plane. By contrast, the posterior keel is lacking.

Remarks. – Originally the anterolateral costae were interpreted as anterior and posterior keel due to the strong torsion of the entire element.

Furnishina sinuata n.sp.
Pl. 15:6, 11–15, 17–20; Fig. 8R

Synonymy. – ☐1959 *Furnishina?* sp. a Müller, p. 454, Pl. 12:13, 16.

Derivation of name. – Latin *sinuatus*, according to the sinuous-like anterior and posterior indentations.

Holotype. – UB 1136 (Pl. 15:17–19).

Type locality. – Stenstorp–Dala.

Type horizon. – *Peltura minor* Zone (Vb).

Material. – 70 specimens.

Occurrence. – Zone III: Grönhögen, Karlsfors, St. Stolan; Zone IV: Nygård; Zone Vb: Stenåsen, Stenstorp–Dala; Zone Vc: Degerhamn, Haggården–Marieberg, S. Möckleby–Degerhamn, Stenstorp–Dala; Zone V undiff.: Brattefors, Degerhamn, Mörbylilla–Albrunna, Ödegården, Stenstorp–Dala, Trolmen.

Diagnosis. – A *Furnishina* with a sinuous-like indentation on both anterior and posterior sides. Anterolateral and posterior costae are broadly rounded.

Description. – Stout, strongly recurved sclerites. The cusp is straight and pointed. Most of the elements are slightly deflected laterally. The anterior side is flattened and may bear a wide, shallow furrow. Contrary to other species, the anterolateral costae are less prominent but appear crudely rounded. The posterior side is convex and projects almost rectangularly from the uppermost part of the base. Both anterior and posterior sides are characterized by a typical insinuation of about the same dimensions. The basal opening extends over more than three-fourths of the entire length, and the basal rim is smooth. The cross-section changes from circular at the apex into subtriangular at the basis. Similar to *F. rara*, one specimen has its pitted basal organ preserved (Pl. 15:12).

Size range. – 310–1520 μm.

Furnishina tortilis (Müller 1959)
Pl. 14:2–19; Fig. 8N, O

Synonymy. – ☐1959 *Scandodus tortilis* n.sp. – Müller, p. 464, Pl. 12:7, 8, 10; Fig. 9. ☐1971 *Scandodus tortilis* Müller – Müller, Pl. 2:1. ☐1979 *Scandodus tortilis* Müller – Bednarczyk, p. 434, Pl. 4:15. ☐1981 *Proscandodus tortilis* (Müller) – Miller *in* Robison, p. W113, Fig. 64.5.

Material. – 1700 specimens.

Occurrence. – Zone I: Backeborg, Gum, Klippan, St.Stolan; Zone II: Ekebacka, Haggården–Marieberg, Österplana, Toreborg; Zone III: Karlsfors, Nygård, St. Stolan; Zone IV: Nygård; Zone Vb: Stenstorp–Dala; Zone Vc: Stenstorp–Dala; Zone V undiff.: Stenstorp–Dala, Stubbegården.

Description. – Slender sclerites characterized by a striking torsion. Recurvature is quite variable. The anterior side is rather flat and may have a broadly rounded costa in its basal section. The posterior side projects basally. It is marked by a median costa which has a certain tunnel-tendency (Pl. 14:10). Due to the strong torsion it appears obliquely upward directed. The anterolateral costae differ in length. That one towards which the median costa is directed is the shorter one. The opposite one is traceable almost up to the tip. On the cusp the cross-section is subtriangular with concave flanks, and passes into an asymmetrical outline at the basis. On some elements the smooth basal rim is extended laterally. The outer surface is smooth, the inner one marked by densely set growth lines. According to the torsion, left and right forms are observable. They occur in approximately equal rates.

On some specimens regarded as variants the posterior side is extremely pronounced and may have small lateral lobes (Pl. 14:14, 16, 18). Another specimen shows pathological growth in having its cusp abruptly bent towards the anterior side (Pl. 14:15).

Size range. – 350–1800 μm.

Comparison. – This form agrees with other *Furnishina* species in the three-costate habit and the paraconodont mode of growth. It is closest to *F. asymmetrica* but differs in the distinct torsion and the relatively narrow basal opening throughout all growth stages. The basal height of *F. tortilis* is reduced in comparison with *F. quadrata*, *F. kleithria* etc.

Remarks. – Based on extremely twisted specimens, *F. tortilis* was originally described as *Scandodus tortilis*. The material on hand gives evidence that the costate edges are homologous to the anterolateral costae rather than to anterior and posterior keels. The supposed lateral opening is in fact the narrow costate posterior side.

Furnishina vasmerae n.sp
Pl. 7:1–4, 6–10, 12, 13, 15, 19, 21; Fig. 8G

Derivation of name. – In honour of Mrs. Marianne Vasmer-Ehses, Bonn.

Holotype. – UB 1016 (Pl. 7:3, 4).

Type locality. – Gum.

Type horizon. – *Agnostus pisiformis* Zone.

Material. – 840 specimens.

Occurrence. – Zone I: Backeborg, Gum, Klippan, St. Stolan; Zone II: Haggården–Marieberg, Toreborg.

Diagnosis. – Asymmetrical sclerites with wing-like anterolateral extensions. Larger specimens have strongly vaulted flanks.

Description. – Slender, asymmetrical cones with a straight to gently recurved cusp. The tip may be slightly uplifted. Lateral deflection is quite a common feature, too. The lower portion of the anterior side is flared and extended to wing-like processes of different length. The posterior side projects almost rectangularly at the base and carries a keel which is traceable up to the tip. The flanks are anterolaterally constricted before vaulting to the basal opening. In small specimens they are little differentiated, only larger ones have highly vaulted sides which do not, however, extend beyond the anterior side. The basal rim is smooth. The outer surface is smooth to faintly annulated, the inner one often has the growth lines exposed (Pl. 7:8).

Size range. – 520–1000 μm.

Comparison. – This form is closest to *F. alata* in the wing-like basal processes. It differs, however, in its strongly asymmetrical outline and the generally larger size. The convex flanks resemble the secondary carinae of *F. bicarinata* except for the posterolateral constrictions. Further, they do not exceed the anterobasal width.

Furnishina sp. indet.
Pl. 13:19, 26

Material. – Figured specimen.

Occurrence. – Zone I: Gum.

Description. – A single specimen which is quite similar to *Furnishina curvata* in its general outline. The keel and one of the anterolateral costae have, however, developed a small denticle. Combined with the different occurrence from *F. curvata* it has to be excluded from that species.

Size. – 310 μm.

Furnishina? sp.
Pl. 14:1

Material. – Figured specimen.

Occurrence. – Zone I: St. Stolan.

Description. – A single specimen with sharp anterolateral costae and a rounded posterior keel which, however, terminates above the basal opening. Contrary to other representatives of the genus, this form carries a small secondary denticle.

Size. – 360 μm.

Genus *Gapparodus* Abaimova 1978
Type species. – *Hertzina? bisulcata* Müller 1959

Abaimova (1978) recognized five species of *Gapparodus*: *G. bellus*, *G. bisulcatus*, *G. bokononi*, *G. heckeri*, and *G. porrectus*. But referring to the generally great variability of Cambrian sclerites, combined with distinctly different cross-sections during growth, we prefer to consider all her species as junior synonyms of *Gapparodus bisulcatus*.

For comparisons, the exact position of the lateral furrows can be recognized with certainty only by studying the basal cross-section of specimens of similar size.

Gapparodus bisulcatus (Müller 1959)
Pl. 3:1–42

Synonymy. – ☐1959 *Hertzina? bisulcata* n.sp. – Müller, p. 456, Pl. 13:22, 23, 27. ☐1966 *Hertzina? bisulcata* Müller – Poulsen, pp. 8–9, Pl. 1:9, text- Fig. 3. ☐1969 *Hertzina bisulcata* Müller – Clark & Robison, p. 1045; Fig. 1d. ☐1969 *Hertzina bisulcata* Müller – Clark & Miller, Fig. 1.2–4. ☐1971 *Hertzina bisulcata* Müller – Müller, pp. 12–13, Pl. 1:7. ☐1973 *Hertzina bisulcata* Müller – Özgül & Gedik, pp. 47–48, Pl. 1, Fig.8. ☐1974 *Hertzina bokononi* n.sp. – Landing, p. 1246; Fig. 1h, i. ☐1974 *Hertzina bisulcata* Müller, with var. 1 & 2 – Landing, p. 1246; Fig. 1c; Figs. 1d–f. ☐1975 *Hertzina bisulcata* Müller – Lee, pp. 80–81, Pl. 1:3; Fig. 2c. ☐1976 *Hertzina? bisulcata* Müller – Bengtson, pp. 191–195, Figs. 5–9. ☐1976 *Hertzina? bisulcata* Müller – Abaimova & Ergaliev, pp. 392–393, Pl. 14:12–15. ☐1978 *Gapparodus heckeri* n.sp. – Abaimova, pp. 79–80, Pl. 7:3, 4; Fig. 1d. ☐1978 *Gapparodus bellus* n.sp. – Abaimova, p. 80, Pl. 7:6, 7; Fig. 1h. ☐1978 *Gapparodus bisulcatus* (Müller) – Abaimova, Pl. 7:8; Fig. 1e. ☐1978 *Gapparodus porrectus* n.sp. – Abaimova, p. 81, Pl. 7:5; Fig. 1u. ☐1987 *Gapparodus bisulcatus* (Müller) – An, pp. 106–107, Pl. 1:6, 8–10, 26.

Material. – 300 specimens.

Occurrence. – Zone I: Backeborg, Gudhem, Gum, Klippan, S. Möckleby; Zone II: Haggården–Marieberg, Österplana, Toreborg, Trolmen; Zone Vb: Stenåsen; Zone V undiff.: Stenstorp–Dala.

Description. – Slender, very long specimens, evenly recurved over their entire length. Anterior and posterior sides are convex; they may differ in width. The flanks are characterized by distinct furrows. During ontogeny they become deeper and gradually shifted towards the posterior side. In general, position as well as the depth of the furrows is quite variable. The basal opening is extremely deep and extends virtually up to the tip. The basal rim is usually broken. The cross-section is circular at the initial part but differentiates fairly early.

There is a remarkable difference in wall-thickness between the sclerites themselves as well as on an individual specimen. Generally the wall is thinner on the posterior side. In our opinion, this feature is a primary character. It might have been caused by duplication of the wall. A common modification is an uneven reduction of the outer layer over the entire anterior side, which may be documented in transverse pseudoribs or patches, respectively (Pl. 3:3, 5).

Size. – Up to 3500 µm.

Genus *Gumella* n.gen.

Derivation of name. – After Gum, Kinnekulle, the main occurrence of that genus.

Type species. – *Gumella cuneata* n.sp.

Diagnosis. – Evenly recurved, multilayered simple cones with deep symmetrical lateral furrows. Reinforcement of the posterior side by an additional clasped unit makes the furrows even more distinct. The somewhat flattened posterior side is often shorter than the anterior one and cuneiformly broken off basally. Therefore the posterolateral edges of the anterior side remain uncovered and display a fibrous structure (Pl. 4:9, 21, 22). The basal opening is extremely deep. In one morphotype the opening splits into a narrow, closed tunnel within the upper half of the element.

Comparison. – The cross-sections of *Gapparodus* and *Gumella* are fairly similar. But in *Gapparodus* the lateral furrows derive from simple depressions, whereas *Gumella* has established an additional stabilizing element in its supplementary posterior unit. Further, the main basal opening runs in the anterior part of the sclerite. By contrast, in *Gapparodus* it is positioned posteriorly due to the thickened anterior side. Furthermore, *Gapparodus* has no polygonal apical structures developed.

Gumella cuneata n.sp.
Pls 4:1–25; 5:1–28

Synonymy. – □1971 *Hertzina elongata* Müller – Müller, Pl. 1:2. □1987 *Gapparodus bisulcatus* (Müller) – An, pp. 106–107, Pl. 1:10.

Derivation of name. – Latin *cuneatus*, cuneiform (wedge-shaped), after the shape of the posterior basis.

Holotype. – UB 980 (Pl. 4:1–5).

Type locality. – Gum.

Type horizon. – *Agnostus pisiformis* Zone.

Material. – 250 alpha + 360 beta specimens.

Occurrence. – Zone I: Backeborg, Gudhem, Gum, Klippan, St. Stolan; Zone II: Haggården–Marieberg, Klippan; Zone Vc: Stenstorp–Dala.

Diagnosis. – As for the genus.

Description. – Evenly proclined simple cones with mostly blunt tips. Partly exfoliated specimens (Pl. 5:5) show polygonal facets on the apical region. They cannot be identified as traces of abrasion as they are restricted to deeper layers. The anterior side is convex, the posterior one flattened. The flanks are characterized by symmetrical posterolateral furrows clasped by an added posterior unit. The overlapping lateral margin is slightly bulgy and makes the furrows

appear particularly deep. The basal opening is exceedingly large and almost reaches the tip. The basal margin is not preserved. From the circular tip the cross-section differentiates fairly early with the insertion of the lateral furrows. The outer side is smooth or marked by 'pseudoribs'. They derive from the former dark opaque outer layer which is only preserved in patches. The light yellow underlying unit is assumed to contain less organic matter. The inner side is smooth.

According to the differentiation of the basal opening, two morphotypes can be distinguished:

Alpha: Apart from the main basal opening, a secondary canal is developed in early growth stages (Pl. 4:1–5). A considerable depth of the lateral furrows on quite a narrow apical part has led to a splitting of the main canal. And before a distinct widening of the flanks, both canals become reunited to a joint basal opening.

Beta: This type converges more evenly towards the apex and the furrows do not considerably affect the inner space (Pl. 5:6, 7). The basal opening remains as single canal with this typical outline from the apex down to the basis.

Size range. – 1650–3450 µm.

Remarks. – Both morphotypes have been found together in 65% of the producing 38 samples. In the largest associations, the alpha:beta ratio varies from 0.7:1 to 1.1:1.

Genus *Hertzina* Müller 1959

Type species. – *Hertzina americana* Müller 1959.

Hertzina elongata Müller 1959
Fig. 9A–K

Synonymy. – □1959 *Hertzina elongata* n.sp. – Müller, pp. 455–456, Pl. 13, Fig.2. □1971 *Hertzina elongata* Müller – Müller, Pl. 1, Fig.3. □1979 *Prooneotodus tenuis* (Müller) – Bednarczyk, pp. 433–434, Pl. 3:6. □1979 *Hertzina elongata* Müller – Bednarczyk, p. 429, Pl. 3:11. □1982 *Hertzina elongata* Müller – An, p. 134, Pl. 10:10a, b. □1984 *Hertzina elongata* Müller – Brasier, pp. 245–246, Fig. 1o–q. □1985 *Hertzina elongata* Müller – Orchard, Pl. 21:2–5.

Material. – 220 specimens.

Occurrence. – Zone I: Backeborg, Gum; Zone II: Ödegården; Zone III: Grönhögen, St. Stolan; Zone IV: Nygård; Zone Va: Dwasiden–Hülsenkrug; Zone Vb: Ödegården, S. Möckleby–Degerhamn, Stenstorp–Dala; Zone Vc: Haggården–Marieberg, Sandtorp, Skår, Smedsgården–Stutagården, Stenstorp–Dala, Trolmen; Zone V undiff.: Degerhamn, Ekeberget, Ödegården, Skår, Stenåsen, Stenstorp–Dala, Tomten, Trolmen.

Description. – Slender, subsymmetrical to more or less asymmetrical cones occurring as left and right forms (Fig. 9B, E). They are gently and evenly recurved. From the pointed apex, the flanks diverge in a low angle. The anterior side is convex, the posterior one flattened to slightly concave. The flanks are rather flat or asymmetrically convex and terminate at posterolateral costae which may show a different

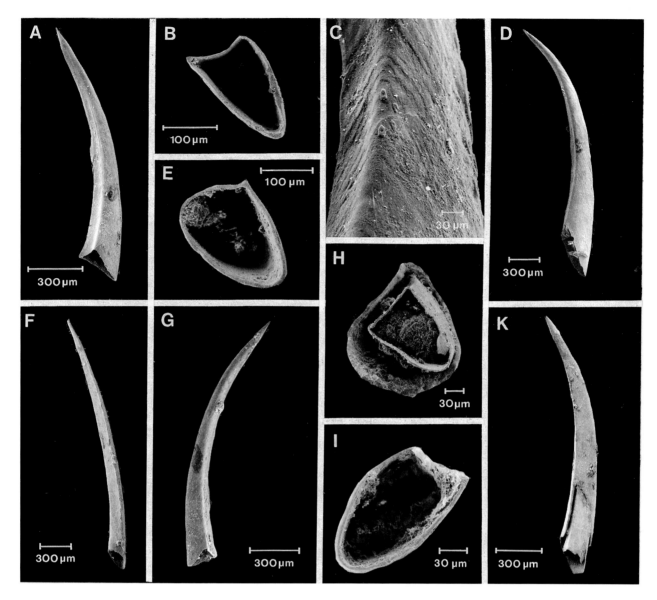

Fig. 9. A–J. *Hertzina elongata* Müller 1959. □A–B. UB 1545 (sample 5659): Degerhamn, zone Vc.Posterolateral and basal view. □C–D. UB 1546 (sample 7218): Möckleby–Degerhamn, zone Vc. Posterolateral and oblique lateral view. Note outcrops of the protoconodontdid growth lines. □E–F. UB 1547 (sample 5951): Stenstorp–Dala, zone Vc. Basal and posterolateral view. □G, I. UB 1548 (sample 1004): Grönhögen, zone Vc. Posterolateral and basal view. □H, K. UB 1549 (sample 5951): Stenstorp–Dala, zone Vc. Basal and posterolateral view of two specimens in 'cone in cone' preservation.

degree of sharpness. They are traceable almost along the entire cone. The basal opening is deep and occupies most of the element. The basal rim is extremely thin and thus usually broken. The cross-section changes from circular apically to semicircular or subtriangular at the basis.

Intraspecific variations include mainly the development of the flanks and posterior side as well as the basal width:height ratio. This resulted in different degrees of symmetry. The specimen on Fig. 9C–D has the edges of the growth lamellae exposed. They are basally directed and may be considered as evidence for a protoconodont mode of growth. Obviously, there are at least two layers of different structure: the inner one is characterized by the outcrops of the growth lines, the outer one by a very faint annulation.

Size range. – 1400–2500 μm.

Genus *Muellerodus* Miller 1980

Synonymy. – □1971 *Muellerina* n.gen. – Szaniawski, p. 407 (non *Muellerina* Bassiouni 1965, an ostracode). □1980 *Muellerodus* n.gen. – Miller, p. 27.

Type species. – *Distacodus*(?) *cambricus* Müller 1959

Remarks. – The genus was established on the basis of specimens with lateral costae and convex anterior and posterior sides. Commonly these characters are combined with a distinct uplifting of the recurved apex. This is, however, not a distinguishing feature on the generic level. It can be observed more or less well-developed on several other taxa, particularly on species of *Furnishina*.

In the following, five species of *Muellerodus* are described: *M. cambricus*, *M. guttulus*, *M.? oelandicus*, *M. pomeranensis*, and *M. subsymmetricus*. They all occur as left and right forms. The basal opening always opens below the point of

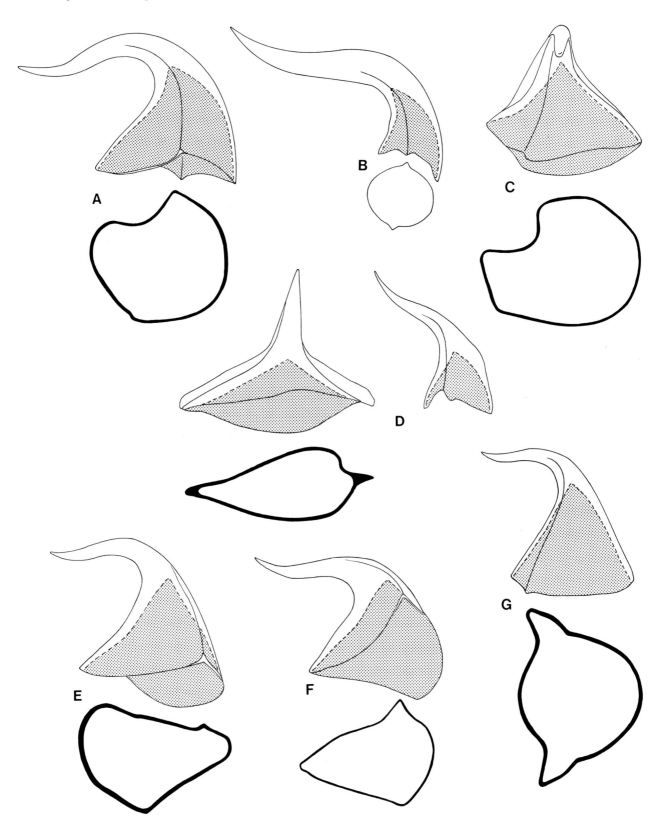

Fig. 10. Outlines of various *Muellerodus* species showing basal openings and cross-sections. □A. *M. cambricus* (Müller 1959). With its slight posterolateral depression it is transitional to *pomeranensis* elements; ×85. □B. *M. cambricus* (Müller 1959); ×135. □C. *M. pomeranensis* Szaniawski 1971; ×65. □D. *M. guttulus* n.sp.; ×155. □E. *M.? oelandicus* (Müller 1959); ×135. □F. *M.? oelandicus* (Müller 1959); ×140. □G. *M. subsymmetricus* n.sp.; ×110.

flexure. The relative length of the cusp decreases from *M. cambricus*, through *M.? oelandicus*, *M. pomeranensis* and *M. subsymmetricus*, to *M. guttulus*. They differ from each other

mainly in the cross-section. *M. cambricus* comprises mostly slender elements with both lateral costae being well-developed. Basal length and height are proportional to each other. *M. guttulus* has an almost tear-shaped basal cross-section. The flanks carry lamina-like costae. *M.? oelandicus* has a variably subtriangular basis. *M. pomeranensis* is characterized by a widely flared base with differently developed costae. *M. subsymmetricus* has posterolateral costae contrary to the other representatives. Further, the posterior side appears somewhat flattened.

Muellerodus cambricus (Müller 1959)
Pl. 18:1–4, 6–13, 15, 17; Fig. 10A, B

Synonymy. – ☐1959 *Distacodus* (?) *cambricus* n.sp. – Müller, p. 450, Pl. 14:1, 2; Fig. 4. ☐1971 *Oneotodus cambricus* (Müller) – Müller, Pl. 2:3. ☐1979 *Muellerina cambrica* (Müller) – Bednarczyk, pp. 429–430, Pl. 2:8. ☐1981 *Muellerodus cambricus* (Müller) – Miller, *in* Robison, p. W112, Fig. 64.3a, b. ☐1982 *Muellerodus pomeranensis* Szaniawski – An, p. 138, Pl. 9:8, 9.

Material. – 520 specimens.

Occurrence. – Zone I: Backeborg, Gössäter, Gum, Klippan, Sätra, S. Möckleby; Zone II: Haggården–Marieberg, Ledsgården, Stubbegården; Zone III: Karlsfors, Nygård; Zone IV: Nygård; Zone Vb: Stenåsen, Stenstorp–Dala; Zone V undiff.: Ödegården, Stenstorp–Dala, Trolmen.

Description. – Slender simple cones with a long, strongly recurved cusp. It may also be slightly twisted and terminate at the pointed, uplifted tip. Anterior and posterior sides are equally convex. Some sclerites have, however, a posterolateral depression, as is typical for *M. pomeranensis* (e.g., Pl. 18:9, 10, 13). They seem to be transitional between both species. Both flanks are characterized by a sharp costa in a somewhat variable position. They are traceable up to at least half the length of the sclerite. Being solid they are restricted to the outer surface and only visible in the inner basal portion, where the wall-thickness decreases considerably (Pl. 18:17). The basal opening is limited to the lower third to half of the unit (Fig. 10A, B). Beyond its top, the costae gradually become indistinct. The basal rim is smooth and gently arched laterally. The circular apical cross-section changes into subcircular or suboval, marked by the short lateral extensions of the costae. The outer surface is almost smooth, the inner one has the paraconodont growth lines exposed (Pl. 18:10).

Intraspecific variation has been observed in position and development of the lateral costae, the length of the cusp and the angle of divergence which may lead to extremely slender specimens with almost parallel sides. Few sclerites have their cusp laterally deflected. The costae are rounded, rather than pointed, and asymmetrically shifted towards the anterior side. Together with a posteriorly expanded base these elements appear fairly asymmetrical (Pl. 18:3).

Size range. – 300–650 μm.

Muellerodus guttulus n.sp.
Pl. 19:1–9, 11; Fig. 10D

Derivation of name. – Adjective formed from Latin *guttula*, small drop, after the crudely tear-shaped basal cross-section.

Holotype. – UB 1180 (Pl. 19:5, 8).

Type locality. – Gum.

Type horizon. – *Agnostus pisiformis* Zone.

Material. – 200 specimens.

Occurrence. – Zone I: Backeborg, Degerhamn, Gössäter, Gum, Haggården–Marieberg, Klippan, Sätra, Trolmen; Zone II: Haggården–Marieberg, Toreborg; Zone Vb: Stenåsen, Stenstorp–Dala; Zone V undiff.: Mörbylilla–Albrunna, Stenstorp–Dala.

Diagnosis. – Paired, very small sclerites with widely flared base. They are broader than long. Pointed lateral costae are well-developed. The basal opening appears approximately tear-shaped.

Description. – Minute asymmetrical specimens, gently recurved at about half length. The cusp is distinctly set off from the widely, laterally flared base. The apex is slightly uplifted. The anterior side is broadly rounded, the shorter, posterior one marked by a posterolateral indentation. The flanks are basally developed as laminae which fade away on the cusp. The basal opening opens below the point of flexure, leaving a cusp which occupies more than half of the entire length (Fig. 10D). The smooth basal rim is usually preserved, as the elements are comparatively thick-walled at the basis. The cross-section changes from circular at the apex into approximately tear-shaped at the basis. The outer surface is smooth, the inner one characterized by coarse growth lamellae.

Size range. – 240–330 μm.

Muellerodus? oelandicus (Müller 1959)
Pl. 20:1–13; Fig. 10E, F

Synonymy. – ☐1959 *Scandodus oelandicus* n.sp. – Müller, p. 463, Pl. 12:14, 15; Fig. 10. ☐1983 *Muellerodus oelandicus* (Müller) s.f. – An *et al.*, p. 109, Pl. 3:9. ☐1986 *Proscandodus oelandicus* (Müller) – Chen & Gong, p. 171, Pl. 34:13; Fig. 67/2.

Material. – 600 specimens.

Occurrence. – Zone I: Backeborg, Gössäter, Gum, Klippan; Zone II: Gössäter, Haggården–Marieberg, Österplana, Stubbegården, Toreborg; Zone III: Grönhögen, Karlsfors, St. Stolan; Zone IV: Nygård; Zone Va: Ödegården; Zone Vb: Grönhögen, Ödegården, Stenstorp–Dala; Zone Vc: Skår, Smedsgården–Stutagården, S. Möckleby–Degerhamn, Trolmen ; Zone V undiff.: Degerhamn, Ekedalen, Ödegården, Rörsberga, Skår, S. Möckleby–Degerhamn, Stenstorp–Dala.

Description. – Simple, strongly asymmetrical, variably twisted, paraconodonts. The cusp is distinctly recurved, with a slight uplifting of the tip. The anterior side is well-rounded, the posterior one is convex–concave. The flanks differ considerably from each other. One of them carries a costa that may also be developed as a laminae. The opposite flank is rounded and not set off from both anterior and posterior sides. The basal opening hardly exceeds the point of flexure and is thus well in accord with the other representatives of the genus (Fig. 10E, F). The basal rim is broadly undulated. The outer surface is marked by a faint annulation, the inner one is smooth.

Size range. – 250–430 μm.

Comparison. – The sclerites are distinguished from *M. cambricus* and *M. pomeranensis* in their general asymmetry and the somewhat variable position of the denticle. Further, they have only one lateral costa developed.

Muellerodus pomeranensis (Szaniawski 1971)
Pl. 18:5, 14, 16, 18–21; Fig. 10C

Synonymy. – ☐1971 *Muellerina pomeranensis* n.sp. – Szaniawski, pp. 408–409, Pls. 1:2; 2:3; 4:1–4. ☐?1979 *Muellerina oelandica* (Müller) – Bednarczyk, p. 430, Pl. 2:7. ☐1981 *Muellerodus pomeranensis* (Szaniawski) – Miller, R.H. *et al.*, Fig. 4I, J. ☐1982 *Muellerodus pomeranensis* (Szaniawski) – An, pp. 138–139, Pls. 9:7; 16:7; 17:1, 3. ☐1983 *Muellerodus pomeranensis* (Szaniawski) – An *et al.*, p. 110, Pl. 3:10. ☐1986 *Muellerodus pomeranensis* (Szaniawski) – Jiang *et al.*, Pl. 2:23. ☐1987 *Muellerodus pomeranensis* (Szaniawski) – An, p. 108, Pl. 3:4, 17.

Material. – 1100 specimens.

Occurrence. – Zone I: Gum; Zone II: Toreborg; Zone III: Karlsfors, Nygård; Zone IV: Grönhögen, Nygård; Zone Va: Ödegården; Zone Vb: Ödegården, Stenåsen, Stenstorp–Dala; Zone Vc: Grönhögen, Gum, Haggården–Marieberg, Kalvene, Sandtorp?, Sellin, Skår, Smedsgården–Stutagården, S. Möckleby–Degerhamn, Trolmen; Zone V undiff.: Brattefors, Ekeberget, Ekedalen, Ödegården, Rörsberga, Skår, Stenstorp–Dala, Tomten, Trolmen.

Description. – Coniform elements with a long and narrow, strongly recurved cusp. It is distinctly set off from the widely flared base. Similar to *M. cambricus* the pointed tip is gently uplifted. The anterior side is broadly rounded, the posterior one is highly convex with a typical depression on one side (Pl. 18:14). Each flank is characterized by a costa which becomes indistinct on the cusp. The one on the depressed side may be developed as a lamina (Pl. 18:20), the opposite one is often indistinct in the basal cross-section. The basal opening is fairly wide and occupies the entire flared lower part of the sclerite (Fig. 10C). The undulated basal rim is smooth. The cross-section passes from circular at the apex into an asymmetrically undulating outline at the basis. The outer surface is smooth, the inner one is marked by the outcrops of the growth lamellae. Variations occur in the width:height ratio of the base and the development of the costae.

Size range. – 330–750 μm.

Muellerodus subsymmetricus n.sp.
Pl. 19:10, 12–19; Fig. 10G

Derivation of name. – Latin *subsymmetricus*, after the basal cross-section.

Holotype. – UB 1186 (Pl. 19:15, 19).

Type locality. – Gum.

Type horizon. – *Agnostus pisiformis* Zone.

Material. – 340 specimens.

Occurrence. – Zone I: Backeborg, Gum, Haggården–Marieberg, Kakeled, Kleva, Klippan, Sätra; Zone II: Haggården–Marieberg, Toreborg.

Diagnosis. – A *Muellerodus* with widely flared base and subsymmetrical posterolateral extensions on either side.

Description. – Above a laterally expanded base the coniform cusp is strongly recurved with a somewhat uplifted tip. A slight torsion of the cusp has been observed on a few specimens. The anterior side is well-rounded, the posterior one is less strongly convex. The blade-like posterolateral costae, which may also appear as laminae, decrease towards the apex; they are traceable over at least two thirds of the entire length. The wide basal opening hardly extends to the point of flexure (Fig. 10G). The basal rim is smooth and even. The cross-section is circular at the apex and changes into a more or less semicircular outline with cornutiform posterolateral extensions. The outer surface is smooth, the inner one marked by the growth lamellae (Pl. 19:19).

Size range. – 280–450 μm.

Genus *Nogamiconus* Miller 1980

Type species. – *Nogamiconus sinensis* (Nogami 1966).

Remarks. – This genus comprises the two species *N. sinensis* (Nogami 1966) and *N. falcifer* n.sp. There is an apparent development from the coniform *Nogamiconus sinensis* to the sickle-shaped *N. falcifer*. An intermediate stage is represented by the rare *N.* sp. A forms which are already horizontally extended and with indicated denticle only. The latter disappears completely in the *falcifer* elements. Our concept excludes *A. cambricus*, *H.*(?) *tricarinata* and *P.*? n.sp. Nogami, assigned to the genus by Miller (1980).

Nogamiconus falcifer n.sp.
Pl. 21:9, 11–23; Fig. 11A

Derivation of name. – Latin *falcifer*, sickle-carrier, after the blade-like upper crest.

Holotype. – UB 1209 (Pl. 21:12, 15).

Type locality. – Trolmen.

Type horizon. – *Peltura scarabaeoides* Zone (Vc).

Material. – 210 alpha + 170 beta specimens.

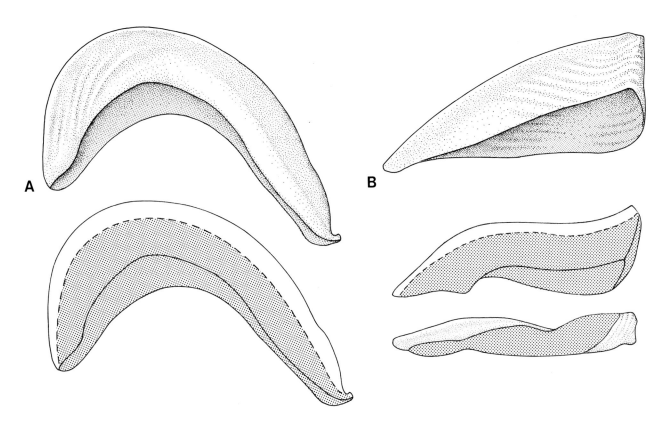

Fig. 11. □A. *Nogamiconus falcifer* n.sp., morphotype alpha. Side view with basal opening; anterior to the left; ×105. □B. *Nogamiconus* sp. Lateral view with basal opening and lower side; anterior to the right; ×85.

Occurrence. – Zone I: Backeborg, Gössäter, Klippan, Ödegården, St. Stolan; Zone II: Degerhamn, Haggården–Marieberg, St. Stolan; Zone III: Grönhögen, St. Stolan; Zone Va?: Ödegården; Zone Vb: Haggården–Marieberg?, Ödegården, Ranstadsverket, S. Möckleby, Stenåsen, Stenstorp–Dala ; Zone Vc: Degerhamn, Ekedalen, Grönhögen, Haggården–Marieberg, Smedsgården–Stutagården?, S. Möckleby–Degerhamn, Trolmen; Zone V undiff.: Brattefors, Ekedalen, Fehmarn, Ödegården, Rörsberga, Skår, Smedsgården–Stutagården, Stenåsen, Stenstorp–Dala, St. Stolan, Stubbegården, Tomten, Uddagården.

Diagnosis. – Comparatively large sclerites with an outline unique among all yet known coniform elements. They are elongate to sickle-shaped with a smooth crest-like upper margin. A denticle is not developed. Two morphotypes are distinguishable. The lower side is taken by a large basal opening. The orientation is analogous to segminiplanate elements defined by Sweet (*in* Robison 1981, p. W15, Fig. 12).

Description. – Elongate to sickle-shaped sclerites which may also be slightly twisted. No denticle is developed. A general character is the upper undenticulated solid crest which decreases towards the anterior and posterior sides. The latter is expanded to a process of variable length. Below the crest the flanks open to a variable degree. The basal opening occupies almost the entire length. As the wall is rather thin and brittle in its lower part, the slightly undulating rim generally is broken. The outer surface is marked by faint ribs which meet the upper margin at an acute angle.

Two morphotypes can be distinguished:

Alpha: Sclerites with convexly curved both upper and lower side resulting in a sickle-shaped appearance. The denticle is completely reduced. Faint ribs, radiating from the posterior third towards the anterior upper margin, serve as a means for orientation (Pl. 21:9). The lower posterior end may be deflected considerably sideways. Anteriorly the flanks may be fused, forming a short keel. Variations are seen in the degree of curvature, torsion and the height of the flanks.

Beta: Straight to convex elements with distinctly differentiated flanks: one is flattened, the other, shorter one is more or less vaulted. Contrary to morphotype alpha, crest and lower rim are not subparallel to each other.

Both types occur in approximately equal numbers. It is in some cases rather difficult to distinguish between right and left specimens because the flanks are subtly differentiated. It is not possible to reconstruct the part of any of the morphotypes within an apparatus. A single cluster shows three specimens of type alpha which are glued together with their posterior parts (Pl. 21:16).

Size range. – 410–1000 µm.

Nogamiconus sinensis (Nogami 1966)

Pl. 21:1–6; Fig. 12A, B

Synonymy. – □1966 *Proacodus? sinensis* n.sp. – Nogami, pp. 356–357, Pl. 10:12–14. □1971 *Proacodus sinensis* Nogami – Müller; Fig. 1c. □1981 *Nogamiconus sinensis* (Nogami) – Miller, *in* Robison, p. W112, Fig. 64.6a, b. □1982 Gen. et sp. indet. – An, p. 156, Pl. 7:7. □1982 *Nogamiconus sinensis*

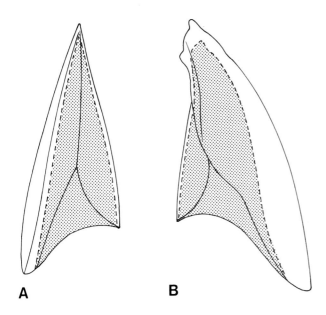

A   **B**

Fig. 12. Nogamiconus sinensis (Nogami 1966). Paired elements with indicated basal openings. Note line of connection particularly on the right sclerite; all ×90.

(Nogami) – An, p. 156, Pl. 8:7. □1983 *Nogamiconus sinensis* (Nogami) – An *et al.*, p. 110, Pl. 30:11. □1986 *Nogamiconus sinensis* (Nogami) – Jiang *et al.*, pp. 48–49, Pl. 2:22

Material. – 30 specimens.

Occurrence. – Zone I: Backeborg, Gum, Klippan; Zone III: Karlsfors; Zone IV: Nygård.

Description. – Subtriangular sclerites with a characteristic growth preference of one flank. The straight pointed apex is usually broken. The anterior side is flattened to slightly concave and bounded by lateral costae. The longer one is a lamina, the other one broadly rounded. The posterior side is angular. The flanks are differently developed: the convex one is folded upon the concave face without achieving a tight closure. The undulating line of connection is distinctly decentralized. The basal opening is very deep and extends up to the tip. The cross-section changes from biconvex at the apex into subtriangular or tear-shaped at the basis. The basal rim is smooth and even. The outer surface is smooth or weakly annulated, the inner one has the paraconodont growth lines exposed (Pl. 21:5)

Size range. – 510–700 µm.

Remarks. – The identity with *Proacodus? sinensis* has been confirmed by Nogami (personal communication). His material also shows a slit on the posterior side.

Nogamiconus sp.
Pl. 21:7, 8, 10; Fig. 11B

Material. – 14 specimens.

Occurrence. – Zone I: St. Stolan; Zone III: St. Stolan; Zone Vb: S. Möckleby; Zone Vc: Ödegården, Stenstorp–Dala; Zone V undiff.: Haggården–Marieberg.

Description. – Very rare; asymmetrical elements with indicated denticle. The latter does not emerge much from the long anterior process. It is developed as a short, sometimes slightly twisted cone which directly passes into the process with evenly converging flanks. The cone may be shifted to one flank which accordingly is concave. The other characters are well in accord with *N. falcifer*.

Size. – About 830 µm.

Genus *Phakelodus* Miller 1984

Remarks. – The type species *Phakelodus tenuis* (Müller 1959) was originally described as *a* new representative of the genus *Oneotodus* Lindström 1954. But at that time, histological differences between proto-, para- and euconodonts had not been studied. In their pioneer work on this topic, Müller & Nogami (1971) distinguished para- and euconodont modes of growth which are radically different from one another. Accordingly, Müller & Nogami added the prefix 'pro-' to all euconodont names erroneously applied to paraconodonts. In this context, *Oneotodus tenuis* changed to *Prooneotodus tenuis*.

A few years later, Bengtson (1976) developed his concept on proto-, para- and euconodonts. He distinguished two types of basal internal accretion and termed the more primitive one with its straight extending lamellae 'protoconodont'. After Szaniawski's (1982, 1983) recognition of such a structure in *Prooneotodus tenuis*, Miller (1984) excluded this form from the paraconodont genus *Prooneotodus*.

In the following, three species are described: *P. elongatus*, *P. simplex* and *P. tenuis*. Both *P. elongatus* and *P. tenuis* are long-ranging and widespread. Although they appear in strikingly similar clusters they are not regarded as morphotypes of a single species. All recognised clusters are unimembrate. Further, the occurrence of these species is indeed overlapping but not consistent enough (Table 3) for them to represent morphotypes.

Phakelodus elongatus (An 1983)
Pl. 1:1–5, 7–9, 12–14, 22

Synonymy. – □1959 *Oneotodus tenuis* n.sp. – Müller, pp. 457–458, Pl. 13:11. □1975 *Prooneotodus tenuis* (Müller) – Lee, pp. 83–84, Pl. 1:16. □?1977 '*Prooneotodus*' *tenuis* (Müller) – Landing, Pl. 1:5. □1979 *Proonoeotdus tenuis* (Müller) – Tipnis & Chatterton, Pl. 29:7–9. □1980 conodont apparatus no. 3 – Abaimova, Fig. 2a, b. □1983 '*Proonoeotodus*' *tenuis* (Müller) – Szaniawski, Figs. 1A, B; 2A–C; 3A–C; 5A; 8. □1982 *Prooneotodus* aff. *tenuis* (Müller) – An, p. 145, Pls. 1:2–6; 2:1, 5. □1983 *Procondontus elongatus* Zhang n.sp. – An *et al.*, p. 125, Pl. 5:4, 5. □1983b *Prooneotodus tenuis* (Müller) – Azmi, pp. 381–382, Pl. 1:5. □1985b *Phakelodus tenuis* (Müller) – Wang, p. 94, Pl. 25:1, 2; Fig. 14/4. □1986 *Phakelodus tenuis* (Müller) – Chen & Gong, Pl. 22:12, 14, Pl. 24:3; Fig. 59.1–3. □1988 *Phakelodus tenuis* (Müller) – Lee, B.S. & Lee, H.Y., Pl. 1:1, 2.

Material. – 620 specimens.

Table 3. Distribution of *Phakelodus elongatus* and *Ph. tenuis* in the various zones.

Zone	number of localities with *Phakelodus*			
	elongatus	*tenuis*	both	total
V	8	5	4	9
Vc	2	5	2	5
Vb	2	3	2	3
Va	–	1	–	1
IV	–	1	–	1
III	1	3	1	3
II	1	7	1	7
I	3	6	3	6

Occurrence. – Zone I: Backeborg, Gum, Klippan; Zone II: Haggården–Marieberg; Zone III: St. Stolan; Zone Vb: Ödegården, Stenstorp–Dala; Zone Vc: Grönhögen, Stenstorp–Dala; Zone V undiff.: Ekeberget, Gum, Ödegården, S. Möckleby, Stenåsen, Stenstorp–Dala, Stubbegården, Trolmen.

Diagnosis. – Slender simple protoconodont sclerites with a tear-shaped cross-section resulting from the keeled posterior side.

Description. – Slender, gently recurved sclerites with a slight deflection to one of the flanks. The anterior side is rounded, the posterior one is keeled almost up to the pointed tip. The basal opening is extremely deep. The cross-section changes from circular apically to tear-shaped at the basis. The latter is quite variable in its width:height ratio, which might also be due to lateral compression. The general length:width ratio varies, too. The basal rim is usually broken. The outer surface is either smooth or characterized by a faint annulation, which may be obliquely crossed by growth lamellae. The inner surface is smooth. Similar to *P. tenuis* these sclerites occur as clusters with up to 26 individuals.

Size range. – 350–800 μm.

Remarks. – This form closely resembles *Protohertzina siciformis* except for the blade-like development of the posterior keel. Further, the other representatives of *Protohertzina* have more differentiated cross-sections with additional posterolateral costae. For this reason we prefer to relate this species to *Phakelodus* rather than to *Protohertzina*. Both *Ph. tenuis* and *Ph. elongatus* have yielded similar clusters while those of *Protohertzina* have not been discovered yet.

Phakelodus simplex n.sp.
Pl. 6:8–15

Synonymy. – □?1973 *Acontiodus* cf. *propinquus* Furnish – Müller, p. 27, Pl. 7:2. □?1981 *Prooneotodus* n.sp. A – Miller, R.H. *et al.*, Fig. 4U–W.

Holotype. – UB 1012 (Pl. 6:14).

Type locality. – Stenstorp–Dala.

Type horizon. – *Peltura scarabaeoides* Zone (Vc).

Material. – 150 specimens.

Occurrence. – Zone I: Gum; Zone II: Haggården–Marieberg, St. Stolan, Toreborg; Zone Vb: Ödegården, Stenstorp–Dala; Zone Vc: Gum, S. Möckleby, S. Möckleby–Degerhamn, Stenstorp–Dala, Trolmen; Zone V undiff.: Degerhamn, Stenstorp–Dala, Trolmen.

Diagnosis. – Stout coniform sclerites with deep basal opening. The large and more or less oval basal cross-section is irregularly flared towards the basis.

Description. – Simple stout elements that are gently recurved, particularly in their apical portion. Both anterior and posterior sides are convex; keels and costae are lacking. The basal opening is extremely deep and extends up to the tip. Thus a proper cusp has not been developed. The cross-section is circular at the apex and rounded to asymmetrically oval at the basis with a variable width:height ratio. The basal rim is smooth and evenly developed. The outer surface may be faintly annulated, the inner one appears smooth. The internal structure is protoconodont and thus well in accord with *Ph. tenuis* and *Ph. elongatus*.

Size range. – 230–600 μm.

Phakelodus tenuis (Müller 1959)
Pls 1:6, 10, 11, 15–21, 23; 2:1–24

Synonymy. – □1959 *Oneotodus tenuis* n.sp. – Müller, p. 457, Pl. 13:13, 14, 20. □1966 *Oneotodus tenuis* Müller– Nogami, p. 356, Pl. 9:11–12. □1971 *Onoeotodus tenuis* Müller – Müller, p. 8, Pl. 1, Figs. 1, 4–6 . □?1973 *Oneotodus tenuis* Müller – Özgül & Gedik, p. 48, Pl. 1:2, 10, 12. □1975 *Prooneotodus tenuis* (Müller) – Lee, pp. 83–84, Pl. 1:14–17; Fig. 2L. □1976 *Prooneotodus tenuis* (Müller) – Miller, R.H. & Paden, p. 596, Pl. 1:20–23. □?1976 *Prooneotodus tenuis* (Müller) – Müller & Andres, pp. 193–200, Pl. 22A, B; Figs. 1a–3. □1977 '*Prooneotodus*' *tenuis* (Müller) – Landing, pp. 1072–1084, Pls. 1:1–4, 6–9; 2:1–11; Fig. 1. □1978 *Prooneotodus tenuis* (Müller) – Abaimova, p. 83, Pl. 8:2, 4. □1980 '*Prooneotodus*' *tenuis* (Müller) – Landing, Ludvigsen & von Bitter, p. 34, Fig. 8 M, N. □1980 conodont apparatuses 1, 2, 4, 5 – Abaimova, Figs. 1a–f; 2c–e. □1981 *Prooneotodus tenuis* (Müller) – An, Pl. 1:16. □1982 *Prooneotodus? tenuis* (Müller) – Szaniawski, pp. 807–810; Fig. 1. □1982 *Prooneotodus tenuis* (Müller) – An, p. 145, Pl. 1:1. □1983b *Prooneotodus tenuis* (Müller) – Azmi, pp. 381–382, Pl. 1:1–4. □1983 '*Prooneotodus*' *tenuis* (Müller) – Landing, pp. 1180–1181, Fig. 10N. □1984 '*Prooneotodus*' *tenuis* (Müller) – Burret & Findlay, Fig. 3d. □1985 '*Prooneotodus*' *tenuis* (Müller) – Orchard, Pl. 2.1:1. □1986 *Phakelodus tenuis* (Müller) – Chen & Gong, pp. 157–158, Pl. 22:7, 8, 12, 14, 19, 21. □1987 *Prooneotodus tenuis* (Müller) – Dong, pp. 169–170, Pl. 1:12; Fig. 3A. □1987 *Prooneotodus tenuis* (Müller) – An, pp. 112–113, Pl. 1:4. □1987 *Phakelodus tenuis* (Müller) – Buggisch & Repetski, p. 158, Pl. 7:a–k. □1987 *Prooneotodus tenuis* (Müller) – Ding, Bao & Cao, p. 80, Pl. 1:17. □1988 *Phakelodus tenuis* (Müller) – Andres, Pl. 1:1–6. □1988 *Phakelodus tenuis* (Müller) – Heredia & Bordonaro, p. 192, Pl. 3:1, 2. □1988 *Phakelodus tenuis* (Müller) – Lee, B.S. & Lee, H.Y., Pl. 1:3.

Material. – 950 specimens.

Occurrence. – Zone I: Backeborg, Degerhamn, Gössäter, Gum, Klippan, St. Stolan; Zone II: Gössäter, Haggården–Marieberg, Ödegården, Österplana, St. Stolan, Stubbegården, Toreborg; Zone III: Grönhögen, Skår, St. Stolan; Zone IV: Nygård; Zone Va?: Ödegården; Zone Vb: Ödegården, Stenåsen, Stenstorp–Dala; Zone Vc: Degerhamn, Grönhögen, S. Möckleby–Degerhamn, Stenstorp–Dala, Trolmen; Zone V undiff.: Brattefors, Ödegården, Stenstorp–Dala, Stubbegården, Trolmen.

Description. – Long, slender, and simple cones. They are gently and evenly recurved. The angle of divergence is exceedingly low. Both anterior and posterior side are convex. The flanks are variably flattened. No keels or costae are developed. The deep basal opening extends nearly or virtually to the entire length. The basal rim has not been preserved in any case but is assumed to have been smooth and evenly developed. The cross-section is circular at the apex and passes into a variably oval outline at the basis. The outer surface is either smooth or marked by oblique transverse ribs, which may form a fishbone structure on the posterior side (Pl. 1:19). On some specimens the transverse sculpture is crossed almost perpendicularly by finer, more densely set ribs forming a cancellate pattern (Pl. 2:4, 5, 18). The edges of the growth lamellae, which are terrace-like where exposed on the outer side (Pl. 1:20), are quite different from this sculpture.

Size range. – 500–2400 µm.

PHAKELODUS *cluster*

Among the protoconodonts, *Phakelodus tenuis* seems to occur most frequently as clusters of unimembrate assemblages. The most complete cluster of *Ph. tenuis* in our collection consists of 17 elements which are bundled in two groups with eight and nine sclerites (Pl. 2:16). The best preserved cluster belongs, however, to *Ph. elongatus* (Pl. 1:2–4, 13) with a total of 26 conodonts arranged in their natural position of a circle instead of the previously assumed two 'half-apparatuses'. Accordingly, the size of the individual elements within an apparatus is quite diverse. Pl. 1:2–4 documents that there were two units of different length with their counterparts, forming a circular trap for nutrition (Pl. 1:13). Such an arrangement is very rarely preserved. Usually the unit is damaged to a variable degree by compaction, so that the single sclerites are shifted against each other. This favours the formation of smaller clusters by breaking apart from larger units. The individual sclerites are glued together with phosphatic matter, which is possible only by a closed position of the apparatus when having caught prey or when resting. For the open position, there is hardly any preservation potential.

Genus *Proacodus* Müller 1959

Type species. – *Proacodus obliquus* Müller 1959

Remarks. – In the original description this monotypic genus comprised elements with both a short and a long lateral process. In the present material these forms occur in 122 samples but together only in 39 ones. Further, the ratio of

the two types in samples with 30–70 specimens is too different to reconstruct an apparatus. As it is very likely that they belong to separate assemblages they have been divided into different species.

Proacodus obliquus Müller 1959
Pl. 22:12–23; Fig. 13C

Synonymy. – □1959 *Proacodus obliquus* n.sp. – Müller, pp. 458–459, Pl. 13:1, 4. □1971 *Proacodus obliquus* Müller – Müller, Pl. 2:2. □1979 *Proacodus obliquus* Müller – Bednarczyk, p. 433, Pl. 1:11, 14. □1981 *Proacodus obliquus* Müller – Miller, *in* Robison, pp. W112–113, Fig. 64.4a, b.

Material. – 550 specimens.

Occurrence. – Zone I: Gum; Zone II: Ödegården; Zone III: Degerhamn; Zone Va: Ödegården; Zone Vb: Haggården–Marieberg, Ödegården, S. Möckleby–Degerhamn?, Stenåsen, Stenstorp–Dala, Tomten ; Zone Vc: Grönhögen, Gum, Kuhbier, Sandtorp?, Skår, S. Möckleby–Degerhamn, Stenstorp–Dala ; Zone V undiff.: Rörsberga, Skår, Smedsgården–Stutagården, Stenstorp–Dala, Trolmen.

Description. – Paired asymmetrical proclined sclerites with a long lateral process as main character. Both anterior and posterior side are broadly rounded. The flank towards which the gently twisted apex points, is convex or even a little expanded on few specimens (Pl. 22:16). The opposite one extends, however, to a keeled process which usually reaches a multiple length of the denticle (Pl. 22:17, 19). It runs in a broad and even, upward directed curvature. The basal opening occupies approximately half the length of the denticle and thus exceeds the point of flexure. It follows the whole lateral process up to the keeled margin (Pl. 22:13). The basal rim is scalloped. The cross-section of the denticle is circular at the apex and rounded to oval at the basis. The outer surface is smooth or faintly annulated.

Size range. – 450–900 µm.

Comparison. – There is an apparent similarity to *Serratocambria minuta* concerning the outer gross morphology. However, neither discrete denticles nor a serration of the process is developed. Also different is the depth of the basal opening within the process: on *Proacodus* it reaches the upper rim, but it is restricted to the lower part on *Serratocambria*. Further, elements of the latter genus are considerably smaller.

Proacodus pulcherus (An 1982)
Pl. 22:3–5, 7–11; Fig. 13A, B

Synonymy. – □1959 *Proacodus obliquus* n.sp. – Müller, pp. 458–459, Pl. 13:2. □1982 *Muellerodus pulcherus* sp.n. – An, p. 139, Pls. 9:13, 15; 10:12, 14.

Material. – 600 specimens.

Occurrence. – Zone II: Berlin, Haggården–Marieberg; Zone III: Degerhamn; Zone Va?: Ödegården; Zone Vb: Ödegården, Stenstorp–Dala; Zone Vc: Degerhamn, Grönhögen, Gum, Haggården–Marieberg, Nästegården, Skår,

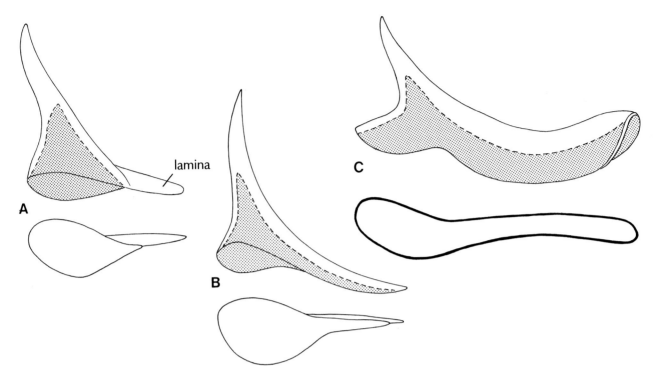

Fig. 13. Outline with basal opening and cross-section of *Proacodus.* □A, B. *P. pulcherus* (An 1982); all ×160. □C. *P. obliquus* Müller 1959; ×125.

S. Möckleby, S. Möckleby–Degerhamn, Stenstorp–Dala, Trolmen; Zone V undiff.: Degerhamn, Kakeled, Öde-gården, S. Möckleby–Degerhamn, Stenstorp–Dala, Stubbe-gården.

Remarks. – We do not follow An's assignment of the form to *Muellerodus* because the latter genus is characterized by distinct lateral costae on both sides. By contrast, *Proacodus* is asymmetrically shaped with one lateral costa.

Description. – Paired slender, proclined specimens with distinctly twisted apex directed either towards the process or away from it. Anterior and posterior side are broadly rounded. One of the flanks is convex and may also be little expanded, the other one continues into a short and straight process. The latter appears as an obliquely downward directed lamina (Pl. 22:5). The keel inserts on the base and gradually widens distally. The depth of the basal opening agrees with *P. obliquus.* The basal rim is smooth and even. The apical cross-section is ciruclar and becomes tear-shaped at the basis. Outer and inner surface are smooth. Variations are seen in the direction of the apex, the length of the lateral process, and the development of the opposite flank.

Size range. – 200–560 μm.

Comparison. – This form differs from *Proacodus obliquus* in its relatively longer denticle and the short lateral process, in some cases developed as mere lamina. The two species are not part of a single taxon's apparatus because their stratigraphic and local co-occurrence is too inconsistent.

Proacodus sp.

Pl. 22:1, 2, 6

Material. – 2 specimens.

Occurrence. – Zone Vc: Grönhögen, S. Möckleby.

Description. – The specimens with the general character of *Proacodus* are marked by a smaller secondary denticle which arises from the lateral process. From a joint basis the basal opening splits into the discrete denticles. Whether these findings represent an exceptional growth or a transformation to multitipped elements is unclear. The rarity may strengthen the former assumption.

Size range. – 380–450 μm.

Genus *Problematoconites* Müller 1959

Type species. – *Problematoconites perforata* Müller 1959.

Remarks. – This genus was established on the basis of numerous holes concentrated on the lower part of the conodont and restricted to the outer layer. The inner surface is dotted with small, densely set pits. The two systems do not correspond with each other and their meaning is still unknown. By contrast, several other genera may be perforated by parasitism, e.g., *Prooneotodus* and *Westergaardodina* as well as *Scolopodus* and *Drepanodus* (Müller & Nogami 1972b). In the latter two cases both element proper and basal plate may be affected. In comparison to other paraconodonts, *Problematoconites* has some sort of 'basal filling' preserved quite frequently. It has a comparatively coarse surface which is marked by numerous small pits of different size and depth (Pl. 23:23). They do, however, not correspond with the outer penetrations.

Problematoconites angustus n.sp.

Pl. 23:11–13, 16, 17

Derivation of name. – Latin *angustus*, narrow.

Holotype. – UB 1251 (Pl. 23:12, 13).

Type locality. – Grönhögen.

Type horizon. – *Peltura scarabaeoides* Zone (Vc).

Material. – 160 specimens.

Occurrence. – Zone II: Berlin, Degerhamn, Stubbegården; Zone III: Grönhögen; Zone Va?: Ödegården; Zone Vb: Ödegården; Zone Vc: Degerhamn, Grönhögen, Gum, Skår, S. Möckleby–Degerhamn, Stenstorp–Dala, Trolmen ; Zone V undiff.: Haggården–Marieberg, Mark Brandenburg, Milltorp, Mörbylilla–Albrunna, Nästegården, Ödegården, Skår, S. Möckleby–Degerhamn, Stenstorp–Dala, Tomten, Trolmen.

Diagnosis. – A *Problematoconites* with narrow flanks which taper to a comparatively long and thin, strongly recurved apex.

Description. – Slender sclerites with a hook-like recurved apex. It is exceedingly thin in comparison with the remainder of the whole element. Anterior and posterior sides are gently rounded, keels and costae are lacking. The flanks may be similarly symmetrically convex to slightly flattened. The basal opening is very large, leaving only one fifth of the entire length for the cusp. The basal rim may be scalloped (Pl. 23:11). The cross-section changes from circular apically to oval at the basis. The outer surface appears smooth to very faintly annulated. The base is variably perforated. For the inner structure see *P. perforata.*

Size range. – 450–1000 µm.

Comparison. – The gross morphology is similar to *Prooneotodus terashimai* which, however, lacks both outer perforations and inner pits. The resemblance is considered as convergency. The species differs from *P. perforata* and *P. asymmetrica* in its slender outline and thin, hook-like, recurved apex.

Problematoconites asymmetricus n.sp.

Pl. 23:21, 23–26

Derivation of name. – Latin *asymmetricus*, after the characteristic posterior differentiation.

Holotype. – UB 1254 (Pl. 23:21, 24).

Type locality. – Grönhögen.

Type horizon. – *Peltura scarabaeoides* Zone (Vc).

Material. – 30 specimens.

Occurrence. – Zone II: Berlin; Zone Va?: Ödegården; Zone Vc: Grönhögen, Nästegården, S. Möckleby–Degerhamn, Trolmen; Zone V undiff.: S. Möckleby–Degerhamn.

Diagnosis. – Paired specimens with the posterior side distinctly differentiated into convex–concave faces. The latter is somewhat expanded and terminates at a lateral costa.

Description. – Gently recurved sclerites. The anterior side is broadly rounded to almost flattened. The convex posterior side becomes, however, distinctly concave on one half from about the midline and continues into an expanded flank terminating at a rounded lateral costa. The opposite flank is convex. The basal rim is scalloped. The outer surface is spread with holes in the lower portion and weakly annulated. For the inner structure see *P. perforata.* From the asymmetrical cross-section, left and right forms are easily distinguished.

Size range. – 900–1300 µm.

Problematoconites perforatus Müller 1959 emend. herein

Pl. 23:1–10, 14, 15, 18–20, 22

Synonymy. – □1959 *Problematoconites perforata* n.sp. – Müller, p. 471, Pl. 15:17. □1971 *Problematoconites perforata* Müller – Druce & Jones, p. 85, Pl. 8:10, 11; Fig. 27. □1971 *Problematoconites perforata* Müller – Müller, p. 12, Pl. 2:11, 13, 14. □1971 *Problematoconites perforata* Müller – Müller & Nogami, p. 14, Pl. 1:1–4; Fig. 1B. □1972b *Problematoconites perforata* Müller – Müller & Nogami, Pl. 14:3, 4; Fig. 1. □1973 *Problematoconites perforata* Müller – Müller, p. 42, Pl. 4:7, 8. □1981 *Problematoconites perforata* Müller – Miller, *in* Robison, p. W113, Fig. 65.4. □1981 *Problematoconites perforata* Müller – An, Pl. 1:12. □1983 *Problematoconites perforata* Müller – An *et al.*, pp. 123–124, Pl. 3:6. □1985b *Problematoconites perforata* Müller – Wang, pp. 94–95, Pl. 21:11, 14. □1986 *Prooneotodus* sp. – Chen & Gong, p. 168, Pl. 17:1; Fig. 64/7. □1986 *Problematoconites perforata* Müller – Jiang *et al.*, Pl. 2:4.

Material. – 420 specimens.

Occurrence. – Zone I: Milltorp; Zone II: Berlin, Degerhamn, Stubbegården; Zone III: Grönhögen, St. Stolan; Zone IV: Nygård; Zone Va: Ödbogården, Ödegården?; Zone Vb: Grönhögen, Ödegården, S. Möckleby, Stenstorp–Dala ; Zone Vc: Degerhamn, Grönhögen, Nästegården, Ödbogården, Skår, S. Möckleby–Degerhamn, Stenåsen, Trolmen; Zone V undiff.: Brattefors, Degerhamn, Haggården–Marieberg, Milltorp, Mörbylilla–Albrunna, Ödegården, Ranstadsverket, Skår, S. Möckleby–Degerhamn, Stenstorp–Dala, Stubbegården, Trolmen.

Description. – Distinctly recurved sclerites with a more or less sickle-shaped outline. The tip is smoothly rounded and does not show traces of abrasion. Costae and keels are lacking. The basal opening is large and occupies more than two thirds of the entire length. The cross-section changes from circular at the apex into variably oval at the basis. The scalloped basal rim is usually preserved. The outer surface is smooth or faintly annulated and dotted with irregularly spread penetrations. They are restricted to the base. Number, shape and arrangement of the holes vary considerably, but this is not defineable taxonomically. On a few elements they show some kind of annular arrangement in being preferentially located between the ribs (Pl. 23:3). The penetrations appear circular to oval or irregularly tear-shaped. They can be observed together on the same sclerite (Pl. 23:7).

Size range. – 300–3000 µm.

Comparison. – see *P. angusta.*

Genus *Prooneotodus* Müller & Nogami 1971

Type species. – *Oneotodus gallatini* Müller 1959.

Prooneotodus gallatini (Müller 1959)
Pl. 24:1–28, ?29

Synonymy. – ☐1959 *Oneotodus gallatini* n.sp. – Müller, p. 457, Pl. 13, Figs. 5–10, 18. ☐1959 *Oneotodus gallatini?* Müller – Müller, Pl. 13:12. ☐1971 *Oneotodus gallatini* Müller – Druce & Jones, p. 81, Pls. 9:9a c; 10:9a–10c; Fig. 26f, g. ☐1973 *Oneotodus gallatini* Müller – Özgül & Gedik, p. 48, Pl. 1:1, 20. ☐?1975 *Oneotodus gallatini* Müller – Lee, pp. 82–83, Pl. 1:2, 12; Fig. 2B, J. ☐1976 *Oneotodus gallatini?* Müller – Abaimova & Ergaliev, pp. 393–394, Pl. 14:11, 17. ☐1980 *Oneotodus gallatini* Müller – Lee, Pl. 1:4. ☐1981 *Prooneotodus gallatini* (Müller) – Miller, *in* Robison, pp. W113–114, Fig. 64.2a–c. ☐1981 *Prooneotodus gallatini* (Müller) – Miller, R.H. *et al.*, Fig. 4K–M. ☐1982 *Prooneotodus gallatini* (Müller) – An, p. 144, Pls. 11:5, 6, 9–14; 16:13. ☐1983b *Prooneotodus gallatini* (Müller) – Azmi, pp. 379–380, Pl. 3:3, 5. ☐1985b *Prooneotodus gallatini* (Müller) – Wang, pp. 96–97, Pl. 21:7–9, 10; Fig. 14/9. ☐1986 '*Prooneotodus*' *gallatini* (Müller) – Chen & Gong, p. 166, Pls. 22:13, 17; 23:2, 3?, 7, 10, 16–18, Pl. 24:12; Fig. 64/2, 5. ☐1986 *Prooneotodus gallatini* (Müller) – Jiang *et al.*, p. 49, Pl. 3:2. ☐1987 *Prooneotodus gallatini* (Müller) – An, p. 112, Pl. 2:1, 2, 5, 6, 11, 15. ☐1987 *Prooneotodus gallatini* (Müller) – Buggisch & Repetski, p. 159, Pl. 8:2. ☐?1988 *Prooneotodus gallatini* (Müller) – Heredia & Bordonaro, p. 193, Pl. 2:4.

Material. – 600 specimens.

Occurrence. – Zone Va: Mark Brandenburg, Ödegården?; Zone Vb: Grönhögen, Ödegården, Stenstorp–Dala; Zone Vc: Degerhamn, Grönhögen, S. Möckleby, S. Möckleby–Degerhamn, Stenstorp–Dala, Trolmen; Zone V undiff.: Kalvene, Milltorp, Ödegården, Skår, Smedsgården–Stutagården, S. Möckleby–Degerhamn, Stenstorp–Dala, Trolmen.

Description. – Symmetrical to asymmetrical sclerites with a broad, even recurvature. The apex is pointed. Both anterior and posterior sides are rounded, the latter may be extended to a narrow process. Keels or costae are lacking. The flanks are convex or flattened, or even concave. The basal opening does not exceed half the entire length. The scalloped basal rim is rarely preserved. The cross-section is circular at the apex and passes into variably oval at the basis. The outer surface is often annulated (Pl. 24, Fig.1), the inner one has the paraconodont growth lines exposed (Pl. 24:17). The different development of the flanks as well as

of the posterior side has led to a differentiation of three intergrading varieties: (a) symmetrical elements with equally convex flanks (Pl. 24:23, 24, 27, 28); (b) asymmetrical sclerites with a convex flank and a flattened or concave counterpart (Pl. 24:3, 7, 19, 20); (c) asymmetrical specimens with a posterior process (Pl. 24:1, 5, 11, 16).

Reference of Pl. 24:29 to the species is somewhat questionable. The big element has a large, circular basis which is distinctly set off from the upper, coniform portion.

Also exceptional is a single specimen which consists of two basally fused cones (Pl. 24:25). Clusters are rather rare (Pl. 24:10, 26).

Size range. – 330–600 µm.

Genus *Prosagittodontus* Müller & Nogami 1971

Type species. – *Sagittodontus dahlmani* Müller 1959.

Prosagittodontus dahlmani (Müller 1959)
Pl. 25:1–22; Fig. 14B

Synonymy. – ☐1959 *Sagittodontus dahlmani* n.sp. – Müller, p. 460, Pl. 14:5, 7, 10; Fig. 8. ☐1971 *Sagittodontus dahlmani* Müller – Müller, Pl. 1:8. ☐1978 *Prosagittodontus dahlmani* (Müller) – Abaimova, pp. 83–84, Pl. 8:3, 5, 7. ☐1981 *Prosagittodontus dahlmani* (Müller) – Miller, *in* Robison, p. W114, Fig. 65.2a, b. ☐1982 *Prosagittodontus dahlmani* (Müller) – An, p. 147, Pl. 5:6. ☐1986 *Prosagittodontus dahlmani* (Müller) – Chen, Zhang & Yu, p. 369 Pl. 2:14, 15. ☐1987 *Prosagittodontus dahlmani* (Müller) – An, p. 113, Pl. 3:23, 24.

Material. – 180 symmetrical, 50 asymmetrical specimens.

Occurrence. – Zone I: Backeborg, Gössäter, Gum; Zone II: Haggården–Marieberg; Zone Vb: Ekeberget, Grönhögen, S. Möckleby–Degerhamn, Stenstorp–Dala; Zone Vc: Degerhamn, Grönhögen, Gum, Karlsro, Kuhbier, S. Möckleby–Degerhamn, Stenåsen, Stenstorp–Dala, Stubbegården, Trolmen; Zone V undiff.: Brattefors, Degerhamn, Ekeberget, Grönhö gen, Ödbogården, Rörsberga, Skår, Stenstorp–Dala, Trolmen.

Description. – Trilobate sclerites that are straight to gently and evenly recurved. The apical angle ranges between 40 and 60°. From the pointed tip the flanks diverge evenly in an almost straight or slightly concave line, separating into distinct lobes. A third, median one, projects posteriorly and is shorter than the lateral ones. The length of the undivided apical part is quite a variable feature. Its relation to the lobes also depends on the size of the whole element. The anterior side is flattened to slightly depressed medially. The flanks are developed as sharp costae. The basal opening is extremely large. The cusp is restricted to the uppermost part of the element. The basal rim is smooth with the typical indentations of both anterior and posterior sides. The degree of indentation is greatest on the anterior side. The posterolateral ones differ in their depth, which leads to a certain subsymmetry. The basal cross-section is crudely

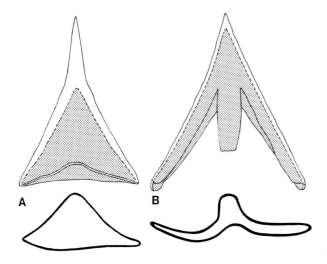

Fig. 14. □A. *Prosagittodontus minimus* n.sp. Outline with basal opening and cross-section; ×105. □B. *P. dahlmani* (Müller 1959). Outline characterised by lobes and indentations and cross-section; ×40.

triangular. The sclerites usually carry broad, transverse ribs. They are obliquely downward directed on the flanks and meet the keeled edge at an acute angle (Pl. 25:8).

Two varieties of *Prosagittodontus dahlmani* have been recognized: (a) subsymmetrical elements with distinctly separated lobes (Pl. 25:8, 16); (b) asymmetrical sclerites that are laterally deflected rather than recurved, so that the apex points towards one of the flanks (Pl. 25:14, 18, 19). This variety is much rarer than the former one. Both, however, yield left and right specimens.

Apart from the apical angle, further variation is seen in the degree of projection of the posterior lobe and the posteriorly curved flanks, which is most distinctive in the basal cross-sections.

Size range. – 240–1300 µm.

Prosagittodontus minimus n.sp.
Pl. 26:10, 12–24; Fig. 14A

Synonymy. – □1959 *Sagittodontus* n.sp. aff. *dunderbergiae* Müller, p. 461, Pl. 14:8.

Holotype. – UB 1295 (Pl. 26:12, 13).

Type locality. – S. Möckleby–Degerhamn.

Type horizon. – *Peltura scarabaeoides* Zone (Vc).

Material. – 650 specimens.

Occurrence. – Zone Va: Ödegården; Zone Vb: Ödegården, S. Möckleby, Stenstorp–Dala; Zone Vc: Degerhamn, Grönhögen, Gum, Sandtorp, Skår, S. Möckleby–Degerhamn, Stenstorp–Dala; Zone V undiff.: Degerhamn, Haggården–Marieberg, Kakeled, Kalvene, Milltorp, Ödegården, S. Möckleby–Degerhamn, Stenåsen, Stenstorp–Dala, Trolmen.

Diagnosis. – A *Prosagittodontus* with comparatively long cusp that is distinctly set off from the base. The latter abruptly widens out in a blunt angle. The flanks are characterized by sharp costae. A posterior, more rounded one, may be present.

Description. – Small, subsymmetrical elements with a long and spine-like, distinctly recurved cusp. The apical angle ranges around 25°. The anterior side is more or less flattened, the posterior one is convex and bears a widely rounded costa that may also become indistinct. The basal opening is somewhat variable in size, but never extends beyond the flared lower part of the cone (Fig. 14A). The basal margin is incompletely preserved but seems to be quite even. Large elements with a well-developed median costa, however, sometimes show a differentiated basal rim with gentle indentations (Pl. 26:20). The circular apical cross-section passes into a crudely triangular one at the basis. The outer surface may be smooth or crossed by faint transverse ribs (Pl. 26:14, 22). The paraconodont growth lines are not exposed on the inner surface.

Variation is observed mainly in length and recurvature of the cusp, the angle of divergence of the base, and the development and position of the posterior costa which may be offcentered to either side causing a certain asymmetry (Pl. 26:16).

Size range. – 300–610 µm.

Comparison. – *Prosagittodontus minimus* differs from *P. dahlmani* in its relatively long, strongly recurved, spine-like cusp. Further, the flanks diverge abruptly rather than evenly. Although *P. minimus* is generally smaller than *P. dahlmani*, it cannot be regarded as the juvenile stage as was suggested by Druce & Jones (1971). Comparison of both types in the same size already reveals considerable differences in outline. A joint occurrence of both species was observed in 35% of the samples. Although *P. minimus* is much more frequent, it was not represented in one sample with 23 specimens of *P. dahlmani*.

Serratocambria n.gen.

Derivation of name. – From Latin *serratus*, after the serrated lateral process.

Type species. – *Serratocambria minuta* n.sp.

Diagnosis. – Minute asymmetrical paraconodonts with free, spine-like cusp and a serrated lateral process. The orientation follows the genus *Proacodus* Müller.

Serratocambria minuta n.sp.
Pl. 27:1–17; Fig. 15A, B

Derivation of name. – Latin *minutus*, after the minute size.

Holotype. – UB 1307 (Pl. 27:10, 11).

Type locality. – Ödegården.

Type horizon. – *Peltura scarabaeoides* Zone (Vc).

Material. – 80 specimens.

Occurrence. – Zone III: St. Stolan; Zone Va: Ödegården, Stenstorp–Dala; Zone Vb: Ödegården?, Stenåsen, Stenstorp–Dala; Zone Vc: Gum, Ödegården, Smedsgården–Stu-

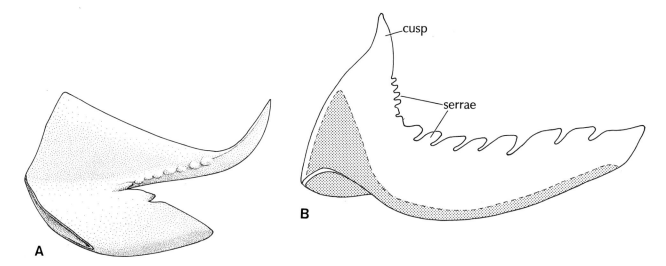

Fig. 15. Serratocambria minuta n.gen., n.sp. □A. Lateral view; ×255. □B. Posterior view with basal opening; ×325.

tagården, Stenstorp–Dala, Trolmen; Zone V undiff.: Degerhamn, Djupadalen, Ekedalen, Ödegården, Stenstorp–Dala.

Diagnosis. – As for the genus.

Description. – Minute asymmetrical sclerites with a strongly recuved, spine-like cusp. It is distinctly set off from the the main denticle. Both anterior and posterior sides are broadly rounded. One of the flanks is convex, the other one extended to a long and compressed, serrated process which lowers down in a broad curvature. It is gently bent opposite to the apex. The serration inserts shortly beneath the small apical part. The massive serrae are laterally directed and become increasingly coarse. They may also flatten so that the free end of the process appears as a mere blade (Pl. 27:8). The depth of the basal opening varies between half of the entire length and almost up to the point of flexure (Fig. 15B). From the main denticle it continues all along the lower part of the process. The basal rim is smooth and may be gently scalloped. The cross-section is circular at the apex and passes into quite a variable outline at the basis of the rounded main denticle with the compressed lateral process. The outer surface is smooth to faintly annulated. The inner one is smooth. The specimens occur as pairs with a ratio of 1:1.3 of left and right elements.

Size range. – 210–270 µm.

Comparison. – See *Proacodus obliquus.*

Trolmenia n.gen.

Derivation of name. – After its main occurence at Trolmen, Kinnekulle.

Type species. – *Trolmenia acies* n.sp.

Diagnosis. – Slender, subsymmetrical, proclined paraconodonts with a short anterior keel. A posterior one may be developed. The gross morphology resembles *Eoconodontus* Miller, which, however, is a conodontophorid.

Trolmenia acies n.sp.
Pl. 26:1–9, 11; Fig. 16A–C.

Derivation of name. – Latin *acies*, keel.

Holotype. – UB 1291 (Pl. 26:4, 9).

Type locality. – Ödegården.

Type horizon. – *Peltura scarabaeoides* Zone (Vc).

Material. – 900 specimens.

Occurrence. – Zone Va: Ödegården; Zone Vb: Ödegården, Smedsgården–Stutagården, S. Möckleby–Degerhamn, Stenstorp–Dala; Zone Vc: Degerhamn, Grönhögen, Gum, Ödegården, Skår, Smedsgården–Stutagården, S. Möckleby–Degerhamn, Stenstorp–Dala, Trolmen; Zone V undiff.: Degerhamn, Ekeberget, Ödegården, Skår, Stenåsen, Stenstorp–Dala, Trolmen.

Diagnosis. – As for the genus.

Description. – Subsymmetrical simple cones that are broadly recurved over the entire length. The cusp is slightly deflected inwardly. The anterior side is characterized by a short keel along the base which may be developed as a lamina. The posterior side usually tapers but does not necessarily terminate in a keel. The flanks are rounded. Shallow depressions or furrows in different positions on either side may be present (Pl. 26:2, 8). The basal opening extends the point of flexure up to half of the entire length (Fig. 16A–C). The remaining cusp is comparatively long and rather narrow. The scalloped basal rim is usually broken. The cross-section changes from tear-shaped to subelliptical or lenticular at the basis. The outer surface is either smooth or annulated, the inner one often has the paraconodont growth lamellae exposed. Presence or absence of a posterior keel is regarded as being subject to intraspecific variation. Left and right elements occur in about the same ratio. Comparable to other paraconodonts, clusters are rather rare. Only a single one, composed of three specimens, is recorded (Pl. 26:11).

Size range. – 440–750 µm.

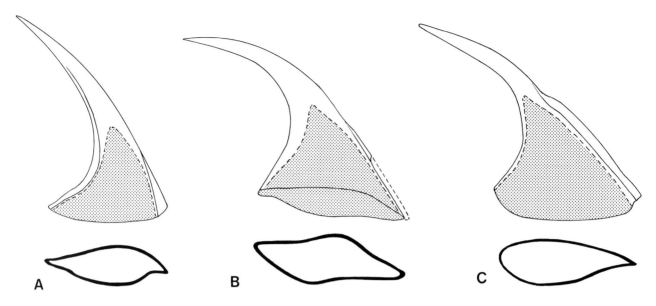

Fig. 16. Variation in *Trolmenia acies* n.gen., n.sp. □A. ×135. □B. ×125. □C. ×145.

Comparison. – Apart from the mere outer resemblance to *Eoconodontus, Trolmenia* is also similar to *Prooneotodus galla-tini* (Müller, 1959). But the generic concept of *Prooneotodus* is based on the absence of keels and costae.

Genus *Westergaardodina* Müller 1959

Type species. – *Westergaardodina bicuspidata* Müller 1959.

Remarks. – As early as 1893 and 1903, Wiman recorded this form as a 'ganz rätselhafter Organismus'. Half a century later, Westergård (1953) described it as Problematicum I. He was followed by Eisenack (1958) who illustrated a U-shaped sclerite which probably also belongs to *Westergaardo-dina* Müller 1959.

With its compound elements, the genus *Westergaardodina* represents a special branch of the Paraconodontida. There are no apparent predecessors which can be linked to this lineage. The existence of many distinctive characters con-tributes to make it the group with the largest number of species among Cambrian conodonts. Concerning the outer gross morphology, two different types can be distin-guished: (a) tricuspidate elements with a comparatively broad and prominent median projection, and (b) bicuspi-date elements with prevalent lateral projections. A median projection, if developed at all in the latter elements, is of subordinate character only. It is much smaller than the lateral ones and thus not relevant for the outer shape. Particularly in this group, right and left elements have been observed.

In general, anterior and posterior sides are developed differently. In most cases the anterior one is smooth and more or less evenly convex. Only in some species does it show characters of taxonomic significance such as furrows or callosities. By contrast, the posterior side is more differ-entiated and accordingly displays the majority of taxonom-ically relevant features, e.g., the development of basal rim, basal and lateral openings and the median keel. Two differ-ent kinds of the latter can be observed: (a) a massive keel

which is linked with the posterior face by a sheet, com-monly shorter than the keel itself, and (b) a hollow keel that continues towards the anterior either in a solid sheet or that is connected with the basal opening by a more or less wide furrow. Some tricuspidate elements lack a definite keel at all. Instead, the median projection is distinctly vaulted as a continuation of the basal opening. In this connection it should be pointed out that the relevant features do not change, but retain their proportions during growth.

An obvious characteristic of the whole group is a great variability of almost all morphological characters. Never-theless, a separation into species is definitely possible be-cause variation is restricted to only certain features which are different for each taxon.

Evolutionary trends. – Dong (1988) described evolutionary developments within *Westergaardodina* as follows: (a) The length/width ratio of the lateral denticles increases. (b) The size of the basal opening decreases.

In our material, Dong's first postulation can be con-firmed only with regard to the increasing appearance of bicuspidate species with minute median projections. Within tricuspidate sclerites, such a trend has not been observed. His second statement, referring to the size of the basal openings, has to be modified in that the open parts of the lateral or basal openings, respectively, decrease. *Westergaardodina amplicava* with comparatively huge contin-uous openings arises already in the lower part of the Upper Cambrian. Later forms, particularly bicuspidate sclerites, usually have their bulging basis closed and exhibit only small to medium-sized lateral openings. Fractures, how-ever, document that the lateral projections may be con-nected by a canal-like structure underneath the bulge. It is unclear whether or not this part was filled with soft tissue during lifetime.

Altogether it does not seem possible to link the species phylogenetically. Accordingly we refrain from establishing subgenera because the relationships between the individual species cannot be traced back with a reasonable degree of confidence. This would require much more knowledge about pre-Late Cambrian representatives than has been achieved by now.

Phosphatic balls. – Bicuspidate elements of various species have been observed with attached phosphatic balls (see Müller 1959, p. 466, Pl. 15:7a). In addition, some sclerites have semicircular notches on the inner side of the lateral projections. They may occur as pairs, leading to a circular structure (Pl. 31:18). Similar to the balls, they take different positions along the flanks and are assumed to be related to the latter. Neither comparable notches nor such balls have been discovered on tricuspidate sclerites.

Westergaardodina ahlbergi n.sp.
Pl. 38:1–11

Synonymy. – □?1973 *Westergaardodina muelleri* Nogami – Müller, p. 48, Pl. 2:9. □1982 *Westergaardodina* cf. *muelleri* Nogami – An, pp. 153–154, Pl. 6:11.

Derivation of name. – In honour of Dr. Per Ahlberg, Lund.

Holotype. – UB 1450 (Pl. 38:1–3).

Type locality. – Gum.

Type horizon. – *Agnostus pisiformis* Zone.

Material. – 100 specimens.

Occurrence. – Zone I: Backeborg, Degerhamn, Gum, St. Stolan; Zone II: Haggården–Marieberg, Toreborg.

Diagnosis. – Medium-sized, tricuspidate elements with short, projecting keel which has the shape of a shoehorn. The turning points are widely rounded and always below the keel. A basal furrow opens to lateral openings.

Description. – Medium-sized, bilaterally symmetrical tricuspidate sclerites. The lateral projections with their outwardly directed tips are exceeded in height by the broad prominent median one. Its hollow posterior keel is rather short and ends above the turning points. The entire unit is fairly straight.

The anterior side is slightly larger than the posterior one and has little surface differentiation except for a shallow median depression. The most prominent feature of the posterior side is a hollow median keel which rises from the tip to about half the length of the element. It projects posteriorly and widens to a somewhat rostrum-like termination. The turning points are surrounded by broad bulgings. In forming quite a strong insinuation they expose part of the the inner surface of the less arched anterior side. The basal opening is well-developed in the median keel. A furrow runs to either side, opening to the slit-like lateral openings. The connection between basal opening and the lateral ones is related to the growth stage of the sclerites. In small specimens it is only indicated between the anterior and posterior sides. In subsequent stages, with the gradual setting apart of both sides, a furrow develops that later forms a continuous opening.

Variable features are the ratio between the lengths of the median and lateral projections, the shape of the latter, and the degree of basal insinuation.

Size range. – 510–650 μm.

Comparison. – See *W. nogamii.*

Remarks. – The species is easily distinguishable from *W. nogamii* because no transition forms have been observed; in addition there is little overlap in their occurrence.

Westergaardodina amplicava Müller 1959
Pl. 33:17–22, Pl. 34:1, 2, 4

Synonymy. – □1959 *Westergaardodina amplicava* n.sp. – Müller, pp. 467–468, Pl. 14:14, 16. □1971 *Westergaardodina amplicava* Müller – Druce & Jones, p. 100, Pl. 8:1, 2; Fig. 15. □1971 *Westergaardodina amplicava* Müller – Jones, Shergold & Druce, Fig. 6. □1971 *Westergaardodina amplicava* Müller – Müller, Pl. 2:7. □1973 *Westergaardodina amplicava* Müller – Müller, p. 46, Pl. 2:10. □1985b *Westergaardodina amplicava* Müller – Wang, pp. 100–101, Pl. 22:7. □1986 *Westergaardodina amplicava* Müller – Chen & Gong, pp. 197–198, Pl. 21:7; Fig. 83. □1987 *Proscandodus oelandicus* (Müller) – Bischoff & Prendergast, p. 51, Fig. 4r.

Material. – 200 specimens.

Occurrence. – Zone II: Berlin, Stubbegården; Zone IV: Nygård; Zone Va: Ödbogården; Zone Vb: Haggården– Marieberg, Ödegården, Stenåsen, Smedsgården–Stutagården, Stenstorp–Dala, Tomten, Trolmen; Zone Vc: Ekeberget, Ekedalen, Grönhögen, Haggården–Marieberg, Nya Dala, Smedsgården–Stutagården?, S. Möckleby– Degerhamn, Stenstorp–Dala, Tomten, Trolmen; Zone V undiff.: Brattefors, Degerhamn, Ekeberget, Nästegården, Ödegården, Ranstadsverket, Skår, Smedsgården–Stutagården, S. Möckleby–Degerhamn, Stenstorp–Dala, Trolmen.

Description. – Large, nearly bilaterally symmetrical, tricuspidate units. The median projection is always smaller than the flaring lateral ones, but all are characterized by hollow keels. The tips of the lateral projections are outward directed. The large basal opening is continuous but only open below.

Contrary to the smooth anterior side, the slightly shorter posterior one has quite strong relief: the median keel is narrow, while the broad asymmetrical lateral ones are somewhat shifted towards the outer margin. Except for the tips, the upper rim is formed by a more or less extended sharp costa. The greatest extent of the basal opening is within the lateral projections. As a typical feature, they are laterally closed.

Size range. – 600–1750 μm.

Westergaardodina auris n.sp.
Pl. 34:13, 14, 17–21

Derivation of name. – Latin *auris*, ear, after the lateral, ear-like widening of the basal opening.

Holotype. – UB 1413 (Pl. 34:13, 14, 17).

Type locality. – Gum.

Type horizon. – *Agnostus pisiformis* Zone.

Material. – 40 specimens.

Occurrence. – Zone I: Gum.

Diagnosis. – Large and compact, tricuspidate sclerites with a strong, solid, posterior keel. The continuous basal opening has ear-like extensions in the lateral projections.

Description. – Fairly massive sclerites with laterally characteristic ear-like widenings of the basal opening. The median projection is smaller to subequal to the lateral ones which have their tips directed outwardly. On large specimens the top end is not acute but extends horizontally (Pl. 34:17).

The anterior side is larger than the posterior one. Its smooth surface may be marginally undulated perpendicular to the rim, a feature more often visible on larger specimens. A shallow longitudinal furrow runs along the median projection. By contrast, the posterior side is strongly differentiated: it is characterized by a solid keel that extends in a broad curve towards the basal rim. Its height increases considerably in the same direction. The broad lateral projections bear pronounced marginal arches at about half length. The marginal rim is reinforced by thickened lamellae that gradually diminish towards the tips; former edges are often visible as terrace-like steps over the entire surface (Pl. 34:21). In the mediobasal plane, the rim is slightly arched, giving view onto the inner surface of the anterior side. The continuous basal opening is extremely large and is deepest within the lateral projections. During growth, anterior and posterior sides became increasingly set apart from each other, which results in a gradual enlargement of the opening.

Variation has been observed in the ratio between the median and lateral projections, the degree of posterolateral arching, and the development of the solid keel, which may also appear somewhat excavated in the lower portion (Pl. 34:20).

Size range. – 950–1400 μm.

Comparison. – See *W. communis* and *W. nogamii.*

Westergaardodina behrae n.sp.
Pl. 37:1–6, 9, 10, 12, 13

Derivation of name. – In honour of Ms. Andrea Behr, Bonn.

Holotype. – UB 1440 (Pl. 37:3, 4, 6).

Type locality. – Stenstorp–Dala.

Type horizon. – *Peltura minor* Zone (Vb).

Material. – 80 specimens.

Occurrence. – Zone Va?: Tomten; Zone Vb: Haggården–Marieberg, Ödegården, S. Möckleby–Degerhamn, Stenåsen, Stenstorp–Dala, Tomten; Zone Vc: Ekedalen, Grönhögen, Skår, S. Möckleby–Degerhamn, Stenstorp–Dala, Tomten; Zone V undiff.: Degerhamn, Smedsgården–Stuta-

gården, S. Möckleby–Degerhamn, Stenåsen, Stenstorp–Dala.

Diagnosis. – Delicate, almost straight tricuspidate units with prominent, keeled median projection. The turning points are deeply incised, the lateral openings are small.

Description. – Small tricuspidate sclerites with prominent median projection. The lateral projections are gently outward-directed and approximately half as wide as the median one. The turning points are deeply incised.

The anterior side has a distinct longitudinal median depression. The surface relief is formed by the somewhat irregular outcrops of the growth lamellae. They are most pronounced around the callous-like turning points. The posterior side carries a median rod-like solid keel. It may continue as a free projecting spur but always ends before the turning points. Anterior and posterior sides are equal in size and fused along the insinuated basal margin. Lateral openings are developed in the upper part of the lateral projections. They are extremely shallow and narrow.

Variation is observed mainly in the relation between median and lateral projections, the length of the keel and the degree of basal insinuation.

Size range. – 100–350 μm.

Comparison. – The main difference from *W. calix* is that *W. behrae* has deeply incised turning points which are located far below the keel. In addition, the lateral projections are much narrower and the lateral openings extend to the tips. Furthermore, *W. behrae* has a much broader median projection in comparison with the lateral ones; in *W. calix* the median projection appears rather narrow. Also the latter is relatively shorter than in *W. behrae.*

Westergaardodina bicuspidata Müller 1959
Pl. 29:13–19

Synonymy. – □1953 Problematicum I – Westergård, Pl. 5:2–4. □1959 *Westergaardodina bicuspidata* n.sp. – Müller, p. 468, Pl. 15:1, 7, 9. □1960 *Westergaardodina bicuspidata* Müller – Müller, Pl. 1:2. □?1971 *Westergaardodina bicuspidata* Müller – Druce & Jones, pp. 100–101, Pl. 7:1–4. □1971 *Westergaardodina bicuspidata* Müller – Müller & Nogami; Fig. 1D. □1972a *Westergaardodina bicuspidata* Müller – Müller & Nogami, Fig. 1B. □1979 *Westergaardodina bicuspidata* Müller – Bednarczyk, p. 435, Pl. 2:2, 13. □?1982 *Westergaardodina bicuspidata* Müller – An, p. 151, Pl. 7:6. □1986 *Westergaardodina bicuspidata* Müller – Chen & Gong, pp. 198–199, Pl. 21:1, 4, 10; Fig. 84.5. □1986 *Westergaardodina bicuspidata* Müller – Jiang *et al.*, p. 53, Pl. 1:17; Fig. 2–12:a. □1987 *Westergaardodina bicuspidata* Müller – An, p. 115, Pl. 3:13.

Material. – 700 specimens.

Occurrence. – Zone I: Gum; Zone II: Berlin, Degerhamn, Fehmarn ; Zone III: Beggerow, Ekedalen, Grönhögen, Karlsfors, Ödegården, S. Möckleby ; Zone IV: Nygård; Zone Va?: Stenstorp–Dala; Zone Vb: Haggården–Marieberg, Ödegården, Smedsgården–Stutagården?, Stenåsen, Stenstorp–Dala, ; Zone Tomten; Zone Vc: Brattefors, Ekedalen, Grönhögen, Gum, Haggården–Marieberg, Ödegården,

Smedsgården–Stutagården, S. Möckleby, Stenstorp–Dala, Tomten, Trolmen; Zone V undiff.: Brattefors, Degerhamn, Ekeberget, Ekedalen, Haggården–Marieberg, Hiddensee, Karlsro, Mark Brandenburg, Milltorp, Mörbylilla–Albrunna, Nästegården, Nya Dala, Ödegården, Rörsberga, Sellin, Skår, S. Möckleby–Degerhamn, Trolmen, Uddagården.

Description. – Medium-sized, subsymmetrical, bicuspidate sclerites with a tiny, keeled, median projection. The lateral projections extend from the subcircular lower part of the element and diverge slightly. The specimens are weakly but evenly recurved.

The anterior side is plain except for a comarginal fold close to the basis. The posterior side is smaller than the anterior one along the lateral openings. In the basal region, a broad bulging raises fairly steeply but tapers off towards the lateral openings. The median projection bears a rod-like keel. Anterior and posterior sides are fused up to a level that generally is above the level of the tip of the median projection. From there, elongate cavities open to either side.

Size range. – 280–710 µm.

Growth development. – On large specimens, the lower, subcircular portion becomes considerably elongated with distinct influence on the general proportions. Also, the lateral projections diverge increasingly during growth.

In the emended definition of *W. bicuspidata*, the species is fairly conservative. Little variation may occur in the general outline, particularly in the lower portion.

Remarks. – Müller (1959, p. 468, Pl. 15:7) noted that increased knowledge might lead to a splitting of the species. We restrict the species to elements closely similar to the holotype, and refer the others (Müller 1959, Pl. 15:1, 9, 10, 14) to the new species *W. polymorpha* and *W. procera*. These species are more slender in comparison with *W. biscuspidata* which has a subcircular lower portion. Contrary to *W. bicuspidata*, *W. procera* has an angular anterior side and *W. polymorpha* is characterised by an ontogenetic increase of the posterior side.

Westergaardodina bohlini Müller 1959
Pl. 29:1–12

Synonymy. – □1953 Problematicum I – Westergård, p. 466, Pl. 5:6, 15. □1959 *Westergaardodina bohlini* n.sp. – Müller, p. 469, Pl. 15:8. □1979 *Westergaardodina bohlini* Müller – Bednarczyk, p. 435, Pl. 2:1. □1987 *Westergaardodina bohlini* Müller – Bischoff & Prendergast, p. 51, Fig. 4i.

Material. – 190 specimens.

Occurrence. – Zone III: Grönhögen, Karlsfors, Ödegården, Nästegården, S. Möckleby; Zone IV: Degerhamn, Karlsfors, ; Zone Va: Ödegården; Zone Vb: Ödegården, Stenstorp–Dala; Zone Vc: Degerhamn, Grönhögen, Haggården–Marieberg, Ödegården?, Smedsgården–Stutagården, Stenstorp–Dala, Trolmen; Zone V undiff.: Degerhamn, Ekedalen, Kalvene, Milltorp, Ödegården, Smedsgården–

Stutagården, Stenåsen, Stenstorp–Dala, St. Stolan, Tomten, Uddagården.

Description. – Medium-sized, symmetrical, bicuspidate sclerites with short but broad, keeled median projection. The turning points are deeply incised. The lower portion of the whole unit is circular to approximately quadrangular; above, the lateral projections diverge. The profile is rather flat.

The anterior side is smooth and has prominent callosities around the narrow turning points followed by a couple of comarginal folds. The median projection is marked by a shallow, longitudinal furrow. The posterior side is slightly smaller than the anterior one. The median projection bears a rounded, rod-like keel, which terminates at a distinct basal bulging. The latter fades away with the opening of the lateral openings. Both anterior and posterior sides are fused along the basis. Lateral openings are large and extend to the top ends of the flanks.

Variation is seen in the general outline, the degree of arching, particularly in the lower portion, the size of the median projection, and the development of the lateral top ends.

There are two comparatively large specimens with flaring lateral projections and closed top ends which are characterized by pointed terminations on either side (Pl. 29:11, 12).

Size range. – 280–750 µm.

Remarks. – In the original description, the cone-like development of the closed lateral top ends had been stressed as a taxonomically significant feature. The enlarged material, however, shows that this character occurs in a comparatively rare variant. A similar development has been observed on a few specimens of *W. bicuspidata*. Therefore, it is considered now as a peculiar feature without taxonomic value.

Westergaardodina calix n.sp.
Pl. 37:7, 8, 11, 14–16

Derivation of name. – Latin *calix*, after the calyx-like outer shape.

Holotype. – UB 1445 (Pl. 37:7, 8).

Type locality. – S. Möckleby–Degerhamn.

Type horizon. – *Peltura scarabaeoides* Zone (Vc).

Material. – 120 specimens.

Occurrence. – Zone Va?: Ödegården; Zone Vb: Brattefors, Haggården–Marieberg, Ödegården, S. Möckleby–Degerhamn, Stenåsen, Stenstorp–Dala; Zone Vc: Ekeberget, Smedsgården–Stutagården, S. Möckleby–Degerhamn, Stenstorp–Dala; Zone V undiff.: Brattefors, Degerhamn, Milltorp, Smedsgården–Stutagården, S. Möckleby–Degerhamn, Stenstorp–Dala, Trolmen.

Diagnosis. – Flat, tricuspidate sclerites of calyx-like appearance. Median projection with rod-like solid keel; relatively small lateral openings.

Description. – Small- to medium-sized, bilaterally symmetrical tricuspidate sclerites with the overall shape of a calyx. In profile they are rather flat. The median projection is equal to or slightly larger than the lateral ones. In relation to the total length, the turning points are fairly high.

The anterior side is smooth with a wide central depression. Slight callosities around the turning points may be developed (Pl. 37:11). The posterior side is a little smaller than the anterior one, particularly along the lateral margins. The median projection bears a distinct rod-like solid keel posteriorly which extends beyond the turning points but does not reach the basal margin. The latter appears rather flattened and is slightly insinuated in the centre. Anterior and posterior sides are fused. They form small openings around the middle of the lateral projections.

There is a certain variation in the depth of the turning points and the development of gentle callosities around them, the size of the lateral openings, and the degree of basal insinuation.

Size range. – 250–570 µm.

Westergaardodina communis n.sp.
Pl. 36:5, 7–17

Synonymy. – □1959 *Westergaardodina tricuspidata* n.sp. – Müller, p. 470, Pl. 15:3, 6. □1971 *Westergaardodina tricuspidata* Müller – Müller, Pl. 2:12. □1982 *Westergaardodina tricuspidata* Müller – An, pp. 155–156, Pl. 7:10. □1983 *Westergaardodina tricuspidata* Müller – An *et al.*, p. 166, Pl. 1:14.

Derivation of name. – Latin *communis*, common, after its common occurrence.

Holotype. – UB 1432 (Pl. 36:10, 13).

Type locality. – Gum.

Type horizon. – *Agnostus pisiformis* Zone.

Material. – 1200 specimens.

Occurrence. – Zone I: Backeborg, Gössäter, Gum, Klippan, St. Stolan, Stubbegården; Zone II: Haggården–Marieberg, Sandtorp, St. Stolan.

Diagnosis. – Tricuspidate units with strongly developed median projection. Often the keel continues as free projecting spur. Lateral openings are well-developed.

Description. – Tricuspidate, fairly large, subsymmetrical sclerites with a gentle posterior bending. The lateral projections extend either in a broad curvature or remain nearly straight. The specimens are characterized by a prominent median posterior projection. Basally the latter constitutes almost double the width of a lateral one.

The anterior side is smooth and may have a median furrow. The posterior side is slightly smaller than the anterior one, which becomes obvious by its median insinuation in the basal region. The large, arrow-like median projection carries a solid keel. Downward from the tip it passes into a spur with a sheet-like connection to the extremely thin posterior side (Pl. 36:7, 14). Usually the distal part of

the spur forms a free projection (Pl. 36:5, 8, 9). The whole structure, however, terminates within the lower third of the element. The area between the keeled structure and outer rim is distinctly concave (Pl. 36:9). Steps of growth lamellae are most pronounced around the turning points. They form U-shaped bulgings which flank the keeled structure and fade away in the upper portion (Pl. 36:17). Having developed sharp edges towards the keel, they leave an area beside and underneath the spur which permits view onto the inner surface of the anterior side.

A proper basal opening is not developed. From the more or less excavated median area, a shallow furrow between the anterior and posterior sides opens laterally to comparatively large cavities.

Variability involves mainly the ratio between the lengths of the median and lateral projections, and the length of the keeled structure. Furthermore, the size of the excavated area obviously is correlated with the development of a free projecting spur.

A mediobasal insinuation of variable degree may be different or equal for anterior and posterior sides. In some cases it may rise above the turning points (Pl. 36:9).

Size range. – 490–1130 µm.

Comparison. – See *W. tricuspidata*.

Westergaardodina concamerata n.sp.
Pl. 34:3, 5–12, 15, 16

Derivation of name. – Latin *con-*, common, and *cameratus*, chambered, after the arched basal margin.

Holotype. – UB 1408 (Pl. 34:5, 11, 12).

Type locality. – Trolmen.

Type horizon. – *Peltura scarabaeoides* Zone (Vc).

Material. – 550 specimens.

Occurrence. – Zone Va?: Ödbogården; Zone Vb: Brattefors, Gum, Haggården–Marieberg, Ödbogården, Stenstorp–Dala; Zone Vc: Grönhögen, Gum, Haggården–Marieberg, Ödbogården, Ranstadsverket, Rörsberga, S. Möckleby, S. Möckleby–Degerhamn, Stenstorp–Dala, Trolmen; Zone V undiff.: Degerhamn, S. Möckleby–Degerhamn, Stenåsen, Stenstorp–Dala, Tomten, Trolmen, Uddagården.

Diagnosis. – Tiny, tricuspidate elements with prominent, keeled median projection. The lateral ones are considerably smaller. There is a pronounced central insinuation of the basis, extending beyond the turning points. A basal opening is absent.

Description. – Small, delicate, symmetrical, tricuspidate sclerites. The pronounced median projection is pointed and keeled. Basally it exceeds double the width of the lateral projections. Nevertheless, the profile of the whole form is fairly flat. The lower rim is strikingly arched in the centre. The lateral projections are narrow and rather short.

The flat anterior side is smooth. On the posterior one, a narrow, tunnel-like, hollow median keel gradually arises from the tip to the basis (Pl. 34:12). A small channel-like connection to the latter may be developed (Pl. 34, Fig.5). The slightly callous-like turning points are deeply incised. They are located below the highest point of the basal insinuation. The equally-sized anterior and posterior sides are fused along the basis. Instead of a basal opening, a small furrow is developed which fades away on the flanks.

Variable features are the ratio between lengths of the median and lateral projections, the growth direction of the latter, as well as the degree of basal insinuation combined with the development of the median keel.

Size range. – 140–230 µm.

Comparison. – See *W. curvata.*

Westergaardodina curvata n.sp.
Pl. 35:1–9, 12

Derivation of name. – Latin *curvatus*, after the typically curved flanks.

Holotype. – UB 1418 (Pl. 35:2, 5, 8).

Type locality. – Grönhögen.

Type horizon. – *Peltura scarabaeoides* Zone (Vc).

Material. – 210 specimens.

Occurrence. – Zone Vb: Brattefors, Haggården–Marieberg, Stenstorp–Dala; Zone Vc: Grönhögen, Gum, Haggården–Marieberg, Ödbogården, Sandtorp?, S. Möckleby–Degerhamn, Stenstorp–Dala, Trolmen; Zone V undiff.: Brattefors, Degerhamn, Ödbogården, Rörsberga, S. Möckleby–Degerhamn, Stubbegården, Trolmen.

Diagnosis. – Small, stout, tricuspidate sclerites. The prominent median projection carries a hollow keel. The basal rim has more or less flat, angular facets. The lateral openings are small.

Description. – Small, subsymmetrical and fairly flat, tricuspidate units. The prominent median projection is pointed and keeled. Nevertheless, the profile is rather flat. Above the turning points the lateral projections are narrower than the median one and diverge distally. Although they are distinctly shorter than the latter, the whole element appears crudely quadrangular. The lower rim with its central insinuation approaches flat, angular facets (Pl. 35:3, 8).

The anterior side is smooth and, apart from a faint median depression, it lacks further differentiation. The posterior side is of equal size and bears a tunnel-like, hollow median keel. It forms the pointed tip of the projection and increases in height towards the basis. From its maximum height the keel continues as a narrow channel to the basis. The turning points are gently callous and located below the greatest height of the keel. Both anterior and posterior sides are basally fused, leaving only a faint but continuous furrow which opens to small openings in the diverging distal part of the lateral projections.

Variation is seen in the shape of median projection, basal opening of the keel, the facets, and the degree of insinuation.

Size range. – 280–450 µm.

Comparison. – This form differs from *W. concamerata* in having the turning points located always above the basal insinuation, in the presence of angular basal facets, and in having lateral openings.

Westergaardodina excentrica n.sp.
Pl. 33:3, 5–16

Derivation of name. – Latin *excentricus*, referring to the position of the median projection.

Holotype. – UB 1392 (Pl. 33:3, 5, 8).

Type locality. – Gum.

Type horizon. – *Agnostus pisiformis* Zone.

Material. – 220 specimens.

Occurrence. – Zone I: Backeborg, Gössäter, Gum, Haggården–Marieberg, Kakeled, Sätra, St. Stolan, Stubbegården, Trolmen; Zone II: Haggården–Marieberg, Klippan, Toreborg.

Diagnosis. – An extremely asymmetrical and twisted bicuspidate *Westergaardodina* in which the positions of median projection and posterior keel do not coincide. The lateral projections are distinctly different from each other. The obliquely incised turning points differ in depth.

Description. – Small, extremely asymmetrical elements with quite an unusual bicuspidate shape. The lateral projections are quite different in outline: the right, smaller one, strongly verges towards the left side; the large, left projection extends horizontally. Both are twisted inwardly. As a striking feature, the median projection is directed towards the left lateral one and hardly emerges from the latter.

The anterior side has a smooth surface. Except for a possible undulation perpendicular to the outer rim (Pl. 33:6) it lacks further differentiation. On the posterior side the median projection is characterized by a hollow keel. It is reduced to a distinct ridge perpendicular to the basis and thus does not reflect the vergency of the projection (Pl. 33:5, 8). The basal opening opens wide towards the straight left lateral projection, the opposite one usually carries a furrow that widens to a flat cavity at the top end.

Variation mainly involves the size of the median projection. In some cases its position is indicated only by the posterior ridge. Other characters are the degree of torsion as well as the width of the lateral projections.

Size range. – 300–530 µm.

Comparison. – *W. excentrica* differs from *W. obliqua* in lacking a distinct tricuspidate outline and a well-developed posterior keel on the large median denticle. The curvature of the whole element is opposite to that of *W. obliqua*, which has a strongly concave anterior side. In addition, *W. excentrica* is characterized by quite a strong torsion – a feature absent in *W. obliqua*. Common to both species is, however, the lack of 'mirror images'.

Westergaardodina latidentata n.sp.
Pl. 35:10, 11, 13–17

Derivation of name. – Latin *latus*, broad, and *dentatus*, dentate, after the large median projection.

Holotype. – UB 1422 (Pl. 35:10, 11).

Type locality. – Stenstorp–Dala.

Type horizon. – *Peltura minor* Zone (Vb).

Material. – 20 specimens.

Occurrence. – Zone Vb: Brattefors, Rörsberga, Stenstorp–Dala; Zone Vc: Grönhögen, Haggården–Marieberg, S. Möckleby–Degerhamn, Stenstorp–Dala, St. Stolan, Trolmen; Zone V undiff.: Trolmen.

Diagnosis. – Tiny, flat, tricuspidate elements with relatively large, blunt median projection. Both anterior and posterior sides are little differentiated. A basal opening is absent.

Description. – Small, compact, symmetrical, tricuspidate sclerites with quite a low relief. The comparatively large median projection has a generally well-rounded top. The lateral projections are narrower than the median one. The basal rim is insinuated.

The anterior side is concave. The turning points have developed gentle callosities which are obliquely inwardly directed (Pl. 35:15, 16). The posterior side is of equal size and slightly convex. It may carry a keel which is, however, restricted to the lower part of the median projection (Pl. 35:13, 17). Instead of a basal opening, a continuous furrow is developed.

Variable characters are the ratio between the lengths of the median and lateral projections, the individual shape of the latter, the development of a posterior keel, the position of the turning points proportional to the entire skeletal length and the degree of basal insinuation.

Size range. – 220–300 µm.

Remarks. – Because of its characteristically developed stout, median projection, this taxon cannot be regarded as representing the small stage of any larger *Westergaardodina* species.

Westergaardodina ligula n.sp.
Pl. 28:1–14

Synonymy. – □1986 *Westergaardodina bicuspidata* Müller – Chen & Gong, pp. 198–199, Pl. 21:15. □?1987 *Westergaardodina bicuspidata* Müller – An, p. 115, Pl. 3:18.

Derivation of name. – Latin *ligula*, spoon.

Holotype. – UB 1311 (Pl. 28:1, 7, 8).

Type locality. – Stenstorp–Dala.

Type horizon. – *Peltura minor* Zone (Vb).

Material. – 200 specimens.

Occurrence. – Zone Va: Ödegården; Zone Vb: Brattefors, Haggården–Marieberg, Ödegården, Stenåsen, Smeds-

gården–Stutagården, S. Möckleby–Degerhamn, Stenstorp–Dala; Zone Vc: Grönhögen, Haggården–Marieberg, Ödegården?, Rörsberga, Smedsgården–Stutagården, S. Möckleby, Stenstorp–Dala, Tomten, Trolmen; Zone V undiff.: Brattefors, Degerhamn, Ekedalen, Fehmarn, Haggården–Marieberg, Milltorp, Mörbylilla–Albrunna, Ödegården, Skår, Stenstorp–Dala, Tomten, Uddagården.

Diagnosis. – Tiny sclerites with extremely small median projection. The basal portion is vaulted like a spoon. The lateral projections are distally slightly diverging; lateral openings are small and shallow.

Description. – Small tricuspidate elements with a minute, spine-like median projection. The lower part of the sclerite is distinctly vaulted towards the anterior; the lateral projections gradually decrease in width and terminate in slightly diverging, rounded tips.

The anterior side is strongly convex and characterized by deeply incised turning points. The tiny median projection, however, makes them appear as a single turning point surrounded by a distinct callosity (Pl. 28:4, 6). Due to the comparatively strong vaulting, the posterior side is somewhat smaller than the anterior one. It is deeply excavated with the maximum depth at about the median projection. Towards the outer rim it is flattened like a brim up to the lateral openings. These open beyond half length of the sclerite and terminate before the tips.

Size range. – 230–500 µm.

Comparison. – In its general outline, the species resembles small stages of *W. polymorpha* and *W. bicuspidata*. It differs, however, from both species in its deeply excavated posterior side which leads to the typical, spoon-like appearance.

Westergaardodina matsushitai Nogami 1966
Pl. 28:15–20

Synonymy. – □1966 *Westergaardodina matsushitai* n.sp. – Nogami, p. 360, Pl. 10:6–8. □1971 *Westergaardodina matsushitai* Nogami – Müller, Fig. 1d. □1978 *Westergaardodina matsushitai* Nogami – Abaimova, p. 86, Pl. 8:6. □1978 *Westergaardodina moessebergensis* Müller – Abaimova, pp. 86–87, Pl. 8:8, ?11. □1981 *Westergaardodina matsushitai* Nogami – An, p. 211, Pl. 1:4. □1982 *Westergaardodina matsushitai* Nogami – An, p. 153, Pl. 6:1–4. □1983 *Westergaardodina matsushitai* Nogami – An *et al.*, pp. 163–164, Pl. 1:3, 4, 8. □1986 *Westergaardodina matsushitai* Nogami – Jiang *et al.*, p. 53, Pl. 1:16; Fig. 12C.

Material. – 70 specimens.

Occurrence. – Zone I: Gum, Backeborg, Haggården–Marieberg, Sätra, Stubbegården; Zone II: Haggården–Marieberg.

Description. – Medium-sized, asymmetrical, bicuspidate elements. A median projection is not developed. The gently diverging lateral projections with rounded top ends are of different length: the left one is always the larger. The basis is obliquely downward directed to the right side. The entire sclerite is only weakly recurved and has an extremely flat profile.

The anterior side is smooth with a distinct depression around the turning point. The posterior side is smaller than the anterior one along nearly the entire outer margin. Both sides are fused at the basis, forming a flat bridge from which large openings develop on either side. As illustrated on Pl. 28:17, the openings are connected internally at the basis.

Size range. – 330–800 μm.

Comparison. – See *W. quadrata.*

Westergaardodina microdentata Zhang 1983
Pl. 40:8–15

Synonymy. – □1983 *Westergaardodina microdentata* Zhang n.sp. – Zhang *in* An *et al.*, p. 164, Pl. 1:15–17; Fig. 9:7. □1986 *Westergaardodina microdentata* Zhang – Jiang *et al.*, Pl. 1:13 (same specimen as in Zhang 1983).

Material. – 110 specimens.

Occurrence. – Zone II: Ödegården; Zone III: S. Möckleby; Zone IV: Degerhamn; Zone Va?: Degerhamn, Ödegården; Zone Vb: Haggården–Marieberg, Ödegården, Smedsgården–Stutagården?, S. Möckleby; Zone Vc: Degerhamn, Haggården–Marieberg, Milltorp, Smedsgården–Stutagården, Stenstorp–Dala, Tomten; Zone V undiff.: Brattefors, Degerhamn, Ekedalen, Mörbylilla–Albrunna, Ödegården, Smedsgården–Stutagården, Stenstorp–Dala, Stubbegården.

Description. – Small, tricuspidate, rounded trapezoidal sclerites with the lateral projections arising above the median one. The profile is rather flattened.

The anterior side exposes pronounced callosities around the turning points. The median projection carries a posterior costa which terminates above the turning points (Pl. 40:8). The posterior side is smaller than the anterior one along the lateral openings. A low median keel emerges from below the tip. From the maximum height in its middle it gradually decreases in both directions and ends up somewhere between turning points and basal margin. Anterior and posterior sides are flattened and fused at the basis. A central insinuation is faintly developed or absent. Lateral openings start above the level of the turning points and terminate before the top ends of the lateral projections.

There is little variation, mostly in the general outline. Differences in the relation between median and lateral projections are generally due to growth, with a considerable enlargement of the lateral ones.

Size range. – 280–600 μm.

Comparison. – See *W. prominens.*

Westergaardodina moessebergensis Müller 1959
Pl. 30:1–8, 10

Synonymy. – □1959 *Westergaardodina moessebergensis* n.sp. – Müller, p. 470, Pl. 14:11, 15. □1966 *Westergaardodina moessebergensis* Müller – Nogami, p. 360, Pl. 10:1, 2. □1971 *Westergaardodina moessebergensis* Müller – Jones, Shergold & Druce, Fig. 6. □1971 *Westergaardodina moessebergensis* Müller – Müller, Pl. 2:5. □?1983 *Westergaardodina moessebergensis* Müller – An *et al.*, pp. 164–165, Pl. 1:5.

Material. – 230 specimens.

Occurrence. – Zone I: Backeborg, Gössäter, Gudhem, Gum, Haggården–Marieberg, Kakeled, Kleva, Klippan, Sätra, St. Stolan, Stubbegården; Zone II: Haggården–Marieberg, Toreborg.

Description. – Medium-sized, asymmetrical, bicuspidate sclerites of quite a compact appearance. A tiny median projection is rarely developed (Müller 1959, Pl. 14:12). The lateral projections differ from each other in length and width and have slightly outward directed tips. The left projection is always the larger one. The turning point is located fairly high. The entire element is only weakly recurved but the crudely bow-shaped posterior side makes the profile rather high.

The flattened posterior side is much smaller than the anterior one with decreasing disparity in size towards the tips. It is asymmetrically concave with the maximum depth at the turning point, tapering off towards the tips. Both sides are mediobasally connected by a narrow bridge. From there, extremely large openings run along the entire lateral projections.

Variation is observed mainly in the development of the lower anterior portion which may be acute, rounded or straight and, influences the lateral outline accordingly.

Size range. – 270–800 μm.

Comparison. – See *W. quadrata.*

Westergaardodina nogamii n.sp.
Pl. 39:1–10

Derivation of name. – In honour of Yasuo Nogami, Kyoto.

Synonymy. – □1953 Problematicum I – Westergård, Pl. 5:8. □1959 *Westergaardodina* n.sp. – Müller, p. 471, Pl. 15:13. □1969 *Westergaardodina muelleri* Nogami – Clark & Robison, p. 1044; Fig. 1c. □1979 *Westergaardodina muelleri* Nogami – Bednarczyk, p. 436, Pl. 2:3, 11.

Holotype. – UB 1459 (Pl. 39:1–3).

Type locality. – Gum.

Type horizon. – *Agnostus pisiformis* Zone.

Material. – 200 specimens.

Occurrence. – Zone I: Backeborg, Degerhamn, Gum, Stubbegården; Zone II: Haggården–Marieberg, St. Stolan, Toreborg; Zone III: Ödegården.

Diagnosis. – Compact, tricuspidate units with prominent hollow keel and continuous basal opening. The lateral projections converge towards the larger median one.

Description. – Large, stout, tricuspidate elements with pronounced median projection. The general outline of the whole form is subcircular. Proportional to the entire length the turning points are fairly high.

The anterior side is weakly convex and lacks distinctive characters. The posterior side is shorter than the anterior one along the entire basal margin. A strong median keel with distinct basal widening develops from the tip down to the basal rim. It terminates far below the turning points. The lateral projections are approximately as broad as the median one and are inwardly curved. The top ends are not acute but slightly flattened. The basal opening is large and continuous to the tips of the lateral projections.

Variation occurs mainly in height and shape of the posterior keel and the level of the turning points, as well as in the size ratio between anterior and posterior sides.

Size range. – 710–1700 μm.

Comparison. – Based on a single specimen, Nogami (1966) erected the new species *W. muelleri*, in which he also included *W.* n.sp. Müller (1959). We refer them to different species because of the differences in gross morphology: *W. muelleri* has much shorter lateral projections with distinctly flattened top ends, and their general divergence contrasts with the converging lateral projections of *W. nogamii*.

W. nogamii differs from *W. ahlbergi* in being much more compact, in the development of the posterior keel, and in the convergence or divergence, respectively, of the lateral projections.

Westergaardodina obliqua Szaniawski 1971
Pl. 33:1, 2, 4

Synonymy. – □1971 *Westergaardodina obliqua* n.sp. – Szaniawski, pp. 410–411, Pl. 1:6, Pl. 5:4–6.

Material. – Figured specimen.

Occurrence. – Zone I: Gum.

Description. – A single specimen can be referred to this species. It fits well into the original description by Szaniawski (1971). Transitions of *W. excentrica* to *W. obliqua* have not been observed.

Size. – 480 μm.

Comparison. – See *W. excentrica*.

Westergaardodina polymorpha n.sp.
Pl. 31:1–21

Synonymy. – □1959 *W. moessebergensis* n.sp. – Müller, p. 470, Pl. 14:12. □1971 *Westergaardodina* cf. *moesserbergensis* Müller – Müller, Pl. 2:6. □1971 *Westergaardodina bicuspidata* Müller – Müller, Pl. 2:8. □?1978 *Westergaardodina?* sp. s.f. – Tipnis, Chatterton & Ludvigsen, Pl. 3:4. □1982 *Westergaardodina bicuspidata* Müller – An, p. 151, Pl. 7:8. □1986

Westergaardodina cf. *moessebergensis* Müller – Kaljo *et al.*, Pl.4:8.

Derivation of name. – Latinized adjective from Greek *polymorphos*, after the pronounced differentiation of shape during growth.

Holotype. – UB 1367 (Pl. 31:13).

Type locality. – S. Möckleby–Degerhamn.

Type horizon. – *Peltura minor* Zone? (Vb?).

Material. – 2500 specimens.

Occurrence. – Zone I: Gum; Zone II: Berlin, Grönhögen, Skår, Stubbegården; Zone IV?: Ödbogården; Zone Vb: Ödegården, Ranstadsverket, Rörsberga, S. Möckleby–Degerhamn?, Stenstorp–Dala, Stoltera, Tomten, Trolmen; Zone Vc: Degerhamn, Grönhögen, Haggården–Marieberg, Karlsfors, Nya Dala, Ödegården, Ranstadsverket, Rörsberga, Smedsgården–Stutagården, S. Möckleby–Degerhamn, Stenstorp–Dala, St. Stolan, Tomten, Trolmen; Zone V undiff.: Brattefors, Degerhamn, Ekeberget, Ekedalen, Grönhögen, Gum, Hiddensee, Kalvene, Karlsro, Kuhbier, Mark Brandenburg, Milltorp, Nästegården, Ödbogården, Ranstadsverket, Rörsberga, Skår, Smedsgården–Stutagården, S. Möckleby–Degerhamn, Stenåsen, Stenstorp–Dala, St. Backor, St. Stolan, Tomten, Trolmen, Uddagården.

Diagnosis. – Large, bicuspidate elements with or without small median projection. The posterior side is much larger than the anterior one. Lateral openings are well-developed.

Description. – Large, symmetrical bicuspidate sclerites. A median projection is either very small or absent. The originally straight lateral projections diverge increasingly during growth. The specimens are only gently recurved, the profile is rather flat.

The anterior side is thin and smooth. There is a depression underneath the median projection which becomes indistinct on large sclerites. The outer margin may become undulated (Pl. 31:8). The median projection is either smooth or has a longitudinal furrow most distinct in the basal portion of the anterior side (Pl. 31:5). The posterior side is of equal size in small specimens but the whole basal portion becomes much increased in later stages (Pl. 31:19). A pronounced bulging at the basis tapers towards the lateral openings. A basal opening is not developed. Lateral ones are comparatively long, extending from the end of the enlarged posterior side up to the tips. They are connected with each other by a shallow furrow on small specimens (Pl. 31:15, 16).

Variation comprises the development of a median projection, the size of the enlarged posterior side, and the degree of lateral divergence.

Size range. – 470–1900 μm.

Comparison. – This species differs from *W. procera* in having a considerably smaller median projection and a much enlarged posterior side on later growth stages. Additionally, diverging lateral projections have not been observed on *W. procera*.

Westergaardodina procera
Pl. 32:1–18

Synonymy. – □1959 *Westergaardodina bicuspidata* n.sp. – Müller, p. 468, Pl. 15:14. □1971 *Westergaardodina bicuspidata* Müller – Müller, Pl. 2:9. □1982 *Westergaardodina fossa* Müller – Fortey, Landing & Skevington, Fig. 8Y.

Derivation of name. – Latin *procerus*, long, slender, after the slim outline of the specimens.

Holotype. – UB 1376 (Pl. 32:1, 2, 6).

Type locality. – Grönhögen.

Type horizon. – *Peltura scarabaeoides* Zone (Vc).

Material. – 830 specimens.

Occurrence. – Zone I: Gum; Zone Va: Tomten; Zone Vb: Gum, Ödegården, Rörsberga, Smedsgården–Stutagården?, S. Möckleby, Stenstorp–Dala, Tomten; Zone Vc: Degerhamn, Grönhögen, Haggården–Marieberg, Nya Dala, Ranstadsverket, Rörsberga, Smedsgården–Stutagården, S. Möckleby–Degerhamn, Stenstorp–Dala, Tomten, Trolmen, Uddagården; Zone V undiff.: Brattefors, Degerhamn, Ekedalen, Kalvene, Ranstadsverket, Skår, S. Möckleby–Degerhamn, Stenåsen, Stenstorp–Dala, Stubbegården, Tomten, Trolmen, Uddagården.

Diagnosis. – Medium-sized, bicuspidate sclerites. The median projection is like a lancet. The anterior side has a distinct basal depression. There are only small and shallow lateral openings.

Description. – Medium-sized, symmetrical, bicuspidate sclerites of quite a delicate appearance. A narrow, median projection may extend over approximately three-fourths of the entire length. In side view, the elements appear somewhat arched. The lateral projections, with their typically concave distal portion, widen out slightly during growth. On a number of specimens, they overlap each other distally, which might indicate quite a high organic content leading to a certain flexibility of the sclerite substance (Pl. 32:1, 2, 7).

The anterior side is gently convex except for the depressed basal portion with the lateral projections forming a blunt angle. The median projection is characterized by a longitudinal furrow which gradually decreases towards the tip (Pl. 32:9, 17). The posterior side is smaller than the anterior one, particularly in the basal region. There a U- or W- shaped bulging is developed which follows the lateral projections and gradually fades away. The median projection bears a broad carina. It nearly takes the entire width leaving only narrow laminae at either side. A basal opening is absent. Instead a hollow space extends within the bulging (Pl. 32:8). Small and shallow lateral openings may open in the concave distal region just beneath the tip.

Variable features include the ratio between the lengths of the median and lateral projections which, however, is also dependent on the size of the whole unit: large ones have comparatively small median projection. The shape of the basis varies between distinctly insinuated and rounded rectangular. This character may also depend on the growth stage.

Size range. – 300–750 μm.

Comparison. – See *W. polymorpha*.

Westergaardodina prominens n.sp.
Pl. 40:1–7

Derivation of name. – Latin *prominens*, after the prominent median projection.

Holotype. – UB 1465 (Pl. 40:2, 3).

Type locality. – Smedsgården–Stutagården.

Type horizon. – *Peltura minor* or *scarabaeoides* Zone (Vb or c).

Material. – 90 specimens.

Occurrence. – Zone III: Grönhögen, Ödegården, S. Möckleby; Zone Va?: Ödegården, Stenstorp–Dala; Zone Vb: Haggården–Marieberg, Ödegården, Rörsberga, Smedsgården–Stutagården? Stenåsen, Stenstorp–Dala; Zone Vc: Milltorp, Smedsgården–Stutagården?, Stenstorp–Dala, Trolmen; Zone V undiff.: Ekedalen, Haggården–Marieberg, Mörbylilla–Albrunna, Ödegården, Smedsgården–Stutagården, Stenstorp–Dala, Tomten.

Diagnosis. – Tiny, tricuspidate elments. Comparatively large median projection, costate and keeled posteriorly. The keel is highest in its middle. The turning points are strongly callous, the lateral openings are small.

Description. – Very small, subsymmetrical, tricuspidate sclerites with strikingly pronounced straight median projection and comparatively short, converging lateral ones which are only about half as wide as the median one. The lower portion of the element is subcircular and appears bowl-shaped.

The anterior side is characterized by a median costa extending from the tip down to about half of the entire length. The turning points are each surrounded by a large callosity (Pl. 40:3). The posterior side bears a solid, rod-like median keel commencing somewhat below the tip and extending below the turning points, not reaching the basal margin. From the maximum height in its middle, the keel tapers in both directions (Pl. 40:7). The turning points are positioned at about half the length of the lateral projections. Particularly in this region, the step-like outcrops of the growth lamellae are well-recognisable (Pl. 40:4, 5). Anterior and posterior sides are equal in size and fused along the basal margin. The latter may be more or less insinuated in the centre. Lateral openings are extremely small and confined to the upper part of the lateral projections.

Variable characters are the relation between median and lateral projections, the size of the callosities, and the degree of basal insinuation.

Size range. – 220–260 μm.

Comparison. – This species is similar to *W. microdentata* in the development of the posterior keel and big callosities on the anterior side. They differ remarkably, however, in the relation of median to lateral projections.

Remarks. – *W. prominens* cannot be regarded as an early growth stage of any of the larger species, in particular because of the peculiar development of the keel. Furthermore, it deviates much in occurrence with those species coming into question.

Westergaardodina quadrata (An 1982)
Pl. 30:9, 11–21

Synonymy. – ☐1982 *Westergaardodina moessebergensis quadrata* subsp. nov. – An, p. 153, Pl. 6:5–8, 10. ☐1983 *Westergaardodina moessebergensis* Müller – An *et al.*, pp. 164–165, Pl. 1:6. ☐1987 *Westergaardodina bicuspidata* Müller – An, p. 115, Pl. 3:13, 18. ☐1990 *Westergaardodina moessebergensis quadrata* An – Dong, Pl. 2:7.

Material. – 4250 specimens.

Occurrence. – Zone I: Backeborg, Gössäter, Gum, Haggården–Marieberg, Kakeled, Klippan, St. Stolan, Stubbegården; Zone II: Haggården–Marieberg.

Description. – Medium-sized, stout, asymmetrical, bicuspidate sclerites. A median projection is lacking. The flared lower part with its narrow turning point is almost circular. From there, the lateral projections diverge increasingly towards the tips. They differ from each other in their proportions, the right one being always larger. The whole element is only little recurved, the profile is not exceedingly pronounced.

The anterior side is smooth with a faint callosity around the turning point. The posterior side is much narrower than the anterior one, particularly in the lower part. The posterior side is strongly concave with its maximum at the turning point and tapers off towards the tip of the lateral projections. Each of the latter bears an ear-like crease in the upper portion (Pl. 30:17). It is more distinct in the narrower projection. The lateral openings are fairly deep, extending to the tips of the lateral projections. Basally they are connceted by a broad furrow.

Variation is rather limited. It includes mainly the ear-like creases on the lateral projections and the flared anterior side.

Growth. – In this form, loss and development of specific characters during growth are fairly distinct. On small stages, the posterior side is only slightly smaller than the anterior one. By contrast, large specimens have a proportionally much smaller posterior side with increasingly raised margins (Pl. 30:14).

Small sclerites have both anterior and posterior side basally fused. During further growth the posterior one becomes bulgy. Subsequently, both sides separate from each other. A broad furrow developed in this way connects the lateral openings (Pl. 30:19).

On the anterior side, small specimens show a distinct callosity around the turning point, which gradually disap-

pears during growth. The ear-like creases evolve that are absent in early stages.

Size range. – 330–700 µm.

Comparison. – *W. quadrata*, *W. moessebergensis* and *W. matsushitai* are similar in the development of asymmetrical lateral projections, a rather flattened anterior side and large lateral openings. *W. quadrata* and *W. moessebergensis* resemble each other in having a strongly concave posterior side. Differences are found mainly in symmetry, the development of the basis, and the position of the turning point. In *W. quadrata*, the right lateral projection is the larger one, contrary to the condition in *W. moessebergensis* and *W. matsushitai*. The latter two forms lack, however, a separation of anterior and posterior side in mature stages. The turning point is located much higher in *W. moessebergensis* than in *W. quadrata* and *W. matsushitai*.

Westergaardodina tricuspidata Müller 1959
Pl. 36:1–4, 6

Synonymy. – ☐1959 *Westergaardodina tricuspidata* n.sp. – Müller, p. 470, Pl. 15:5, 6. ☐1965 *Westergaardodina tricuspidata* Müller – Grant, p. 145, Pl. 15:31. ☐1979 *Westergaardodina tricuspidata* Müller – Bednarczyk, pp. 436–437, Pl. 2:5.

Material. – 50 specimens.

Occurrence. – Zone II: Toreborg; Zone III: Grönhögen, Ödegården; Zone V undiff.: Mörbylilla–Albrunna, Trolmen.

Description. – Medium-sized tricuspidate sclerites with prominent median projection which always exceeds the lateral ones in length. Combined with a mediobasal insinuation, the converging lateral projections make the whole unit appear inverted 'heart-shaped'. In side view it is almost straight (Pl. 36:6).

The anterior side is highly differentiated: the median projection is characterized by a broad central furrow which gradually decreases towards the tip. The deeply incised turning points are surrounded by rows of prominent callosities which are developed as comarginal folds (Pl. 36:3). The posterior side is a little smaller than the anterior one which results in a shallow continuous furrow. A median posterior keel of variable length may bear a short spur near its apical end. Underneath and in its continuation, a narrow furrow extends to the basis. A striking feature are U-shaped bulgings around the turning points with their maximum dimension in the basal part. The lateral openings are rather small and restricted to the upper part of the flanks.

Size range. – 370–650 µm.

Comparison. – This species differs from *W. communis* in the development of the anterior side with strong callosities and comarginal folds, the converging lateral projections, and smaller lateral openings. A proper basal opening is absent.

Remarks. – The original material of Müller 1959 has been substantially increased, which provides more detailed information about the horizontal and vertical distribution of the individual taxa. The forms closely resembling the holotype (pl. 15:5) have been observed in zones III and V, except for a single specimen from zone II at Toreborg. The other sclerites, referable to Müller 1959, Pl. 15:3, 6, have been split off as the separate species *W. communis* based on morphological features and their occurrences restricted to zones I and II. It is likely that *W. tricuspidata* s.str. was restricted to a facies with current-aerated water, while *W. communis*, particularly of zone I, occured in oxygen-poor, stillwater sediments.

Westergaardodina wimani Szaniawski 1971

Pl. 28:21–27

Synonymy. – □1971 *Westergaardodina wimani* n.sp. – Szaniawski, pp. 409–410, Pls. 1:5; 5:7–9. □1982 *Westergaardodina wimani* Szaniawski – An, p. 156, Pl. 9:9. □1982 *Westergaardodina semitricuspidata* n.sp. – An, p. 155, Pl. 7:11, 12. □1986 *Westergaardodina wimani* Szaniawski – Jiang *et al.*, Pl. 1:14 (same specimen as in An 1982).

Material. – 560 specimens.

Occurrence. – Zone I: Backeborg, Gum, Haggården–Marieberg, Kakeled, Kleva, Klippan, Sätra, St. Stolan, Stubbegården; Zone II: Degerhamn, Grönhögen, Haggården–Marieberg, Ledsgården, Sandtorp, St. Stolan, Toreborg.

Description. – A slender, medium-sized, bilaterally asymmetrical bicuspidate *Westergaardodina*. The lateral projections differ considerably in length and are straight to slightly diverging. Right and left elements occur as mirror images. The profile is flat except for a slight bend in the lower portion.

    The anterior side is smooth and lacks distinctive characters. Only a minor callosity is developed around the narrow, deeply incised turning point. The posterior side is smaller than the anterior one, particularly along the larger lateral projection. The basis is characterized by a distinct bulging which diminishes on the lateral projections. Both sides are fused basally. Shallow lateral openings are differently developed: an opening extends over most part of the larger side and may continue to the basis as a furrow. On the shorter projection it is missing or weakly developed within the upper part.

Size range. – 630–900 µm.

Comparison. – There is a total of five known asymmetrical bicuspidate species: *W. excentrica*, *W. matsushitai*, *W. moessebergensis*, *W. quadrata* and *W. wimani*. Apart from obvious morphological differences, the latter species is distinct from all others by its paired occurrence.

Order Conodontophorida Eichenberg 1930

Genus *Acodus* Pander 1856

Type species. – *Acodus erectus* Ulrich & Bassler 1926.

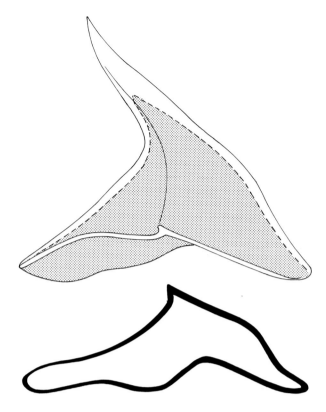

Fig. 17. Acodus cambricus Nogami 1966. Lateral view with basal cavity and cross-section; ×310.

Acodus cambricus Nogami 1967

Pl. 12:20, 21; Fig. 17

Synonymy. – □1967 *Acodus cambricus* n.sp. – Nogami, pp. 213–214, Pl. 1:1–4; Fig. 1. □1987 '*Acodus*' *cambricus* Nogami – An, p. 104, Pl. 2:22.

Material. – 160 specimens.

Occurrence. – Zone I: Backeborg, Gum, Klippan, Sätra; Zone II: Haggården–Marieberg, Stubbegården; Zone Vb: Stenstorp–Dala; Zone Vc: S. Möckleby–Degerhamn, Stenstorp–Dala; Zone V undiff.: Stenstorp–Dala, Stubbegården.

Description. – Small specimens, distinctly recurved and deflected inwardly. Sometimes the pointed apex is even slightly arched upwards. Anterior and posterior sides are rounded costate. One of the flanks, designated as outer one, carries a sharp costae that extends from the cusp to the basal rim. With its broad carina the opposite flank appears undulated. The basal cavity extends up to two thirds of the entire length (Fig. 17). The basal rim is usually broken. The cross-section passes from circular at the apex into roughly triangular at the basal margin.

Size range. – 280–340 µm.

Cambropustula n.gen.

Derivation of name. – From Latin *pustula*, bubble, after the pustulose surface.

Type species. – *Cambropustula kinnekullensis* n.sp.

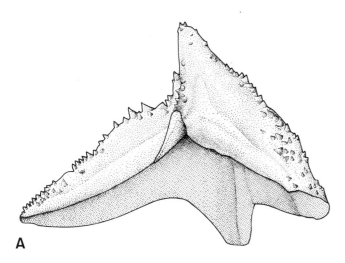

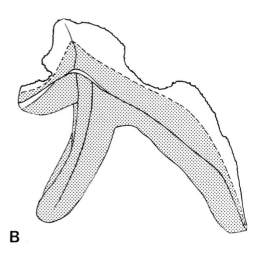

A   B

Fig. 18. Cambropustula kinnekullensis n.gen., n.sp. □A. Outline in oblique posterior view; ×125 □B. Oblique lateral view with basal cavity (reconstruction from a thin-section); ×125.

Diagnosis. – Small, asymmetrical, ramiform to pectiniform, multimembrate euconodonts. They are composed of three to four processes extending from a main cusp. One of the morphotypes lacks such a cusp. A main character are the densely and irregularly set pustules or even denticles forming continuous bands along the crests of the processes. The large basal cavity takes the entire underside almost up to the tip.

Comparison. – *Polonodous* Dzik (1976) is a Middle Ordovician multimembrate genus that comprises polyplacognathiform and ambalodontiform quadrilobate elements. The lobes may split into further, secondary ones. The tubercles of the upper surface are arranged in concentric and radial rows. The similarity to *Cambropustula* is rather superficial, because the latter has no divided lobes and the arrangement of the pustules is still too indifferent and no species character. Nevertheless, a convergency between *Cambropustula kinnekullensis* morphotype delta and *Polonodus* is apparent.

The multielement Lower Ordovician species *Fryxellodontus? corbatoi* Serpagli 1974 [=*Polonodus?* corbatoi (Serpagli) of Stouge & Bagnoli 1988] and *Fryxellodontus? ruedemanni* Landing 1976 are similar to *Cambropustula* in having three processes branching off from the main cusp and in the tendency to form nodes or denticles on their crests. Differences from *Cambropustula* involve in particular the outer shape and mode of ornamentation. In view of the considerable stratigraphic gap between these forms, a close relationship is difficult to demonstrate.

Cambropustula kinnekullensis n.sp.

Pls 44:1–21; 45:1–21; Figs. 18A, B, 19A–D

Derivation of name. – After its occurrence at the Kinnekulle.

Holotype. – UB 1538 (Pl. 45:11, 14).

Type locality. – Gum.

Type horizon. – *Agnostus pisiformis* Zone.

Material. – 60 alpha, 20 beta, 45 gamma, 25 delta elements.

Occurrence. – Zone I: Backeborg, Gum; Zone II: Haggården–Marieberg.

Diagnosis. – As for the genus.

Description. – Proclined, highly differentiated, multimembrate sclerites which occur as left and right forms (Fig. 19). The orientation of the elements within this genus is difficult and may be somewhat arbitrary. It is explained separately for every morphotype. The basal cavity is extremely large. The basal rim is even and may appear somewhat thickened. The sculpture consists of pustules or even denticles topping the crest; a few may also occur above the basal rim. The internal structure is euconodont as has been proved on orientated thin-sections and etched fractures. As *Cambropustula* is restricted to the lower Upper Cambrian, it is the oldest conodontophorid with such an advanced shape. According to its outline, four morphotypes can be distinguished:

Alpha: Subsymmetrical, alate ramiform elements with a gently recurved apex. This apex may be expressed as a comparatively large denticle (Pl. 44:3, 6), a granular node (Pl. 44:10), or as an acute top of the pustulose crest (Pl. 44:4, 5). The flanks are extended and wing-like. The anterior side is convex and marked by a shallow median depression (Pl. 44:4). The concave posterior side bears a process which is smaller than the lateral ones; it appears either as a rib or an expanded lobe (Pl. 44:2).

Beta: This form is characterised by strong differential growth. The lateral processes are much inequal in length. The shorter one is of about the same size as the posterior process. The posterior side has been designated in accordance to the vergency of the largest process (Fig. 19D).

Types gamma and delta are quadrilobate.

Gamma: These elements approach the pectiniform type although they are still similar to type alpha. The additional process originates anteriorly from the former median depression.

Delta: This is the most advanced type. Its strongly asymmetrical outline is characterized by four processes. The apex is reduced almost down to the somewhat flattened

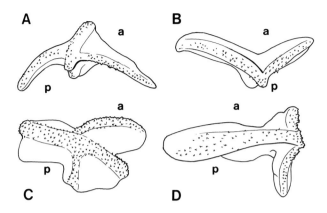

Fig. 19. Orientation in *Cambropustula.* □A. Morphotype gamma □B. Morphotype alpha. □C. Morphotype delta. □D. Morphotype beta.

crest which is entirely covered by a broad band of densely set pustules (Pl. 45:15, 18).

Size range. – 290–460 μm.

Genus *Coelocerodontus* Ethington 1959

Synonymy. – □1987 *Diaphanodus* n.gen. – Bagnoli, Barnes & Stevens, p. 155.

Type species. – *Coelocerodontus trigonius* Ethington 1959.

Remarks. – This genus comprises thin-walled simple cones with an extremely large basal cavity. In many cases this may have led to lateral compression all the way to completely collapsed elements. Thus the original cross-section is difficult to reconstruct.

In the *Treatise on Invertebrate Paleontology*, Klapper & Bergström (*in* Robison 1981, p. W141) referred the genus to the Conodontophorida and included it in the family Belodellidae. Bagnoli, Barnes & Stevens (1987, p. 155) excluded *Coelocerodontus latus* and *C. variabilis* from this genus. They compared these species with paraconodonts due to the lack of white matter and the presence of transverse growth lines. This conclusion cannot be confirmed with our material. Many of our specimens show a faint annulation which cannot be interpretated as growth lamellae. Landing (1983, p. 1172) suggested, that they are '...not euconodonts and may be proto- or paraconodonts histologically'. Since the internal structure has not been studied in detail, corrections on the suprageneric level should be avoided.

As has been documented by Andres (1988, Pl. 9:3–6), *C. latus* and *C. bicostatus* occur in the same cluster and accordingly have to be regarded as morphotypes within an apparatus. Both species have been united under the name *C. bicostatus.*

Coelocerodontus bicostatus van Wamel 1974
Pl. 41:1–21; Fig. 20A–D

Synonymy. – □1974 *Coelocerodontus bicostatus* van Wamel, pp. 55–56, Pl. 3:2. □1974 *Coelocerodontus latus* n.sp. – van Wamel, pp. 56–57. Pl. 1:2. □1983 *Coelocerodontus? bicostatus* van Wamel – Landing, pp. 1172–1173, Fig. 10A, B. □1984 *Vanwamelodus latus* [nomen nudum] – Apollonov, Chugaeva & Dubinina, Pl. 27:25. □1984 *Vanwamelodus* [nomen nudum] sp. 1 – Apollonov, Chugaeva & Dubinina, Pl. 27:31, 33. □?1984 *Vanwamelodus* [nomen nudum] sp. 2 – Apollonov, Chugaeva & Dubinina, Pl. 27:28. □1984 *Vanwamelodus* [nomen nudum] sp. 3 – Apollonov, Chugaeva & Dubinina, Pl. 27:30. □1985b *Nogamiconus* sp. – Wang, p. 94, Pl. 22:2, 5, 6; Fig. 14/12. □1986 *Stenodontus jilingensis* n.sp. – Chen & Gong, pp. 187–188, Pls. 18:2, 4–7, 9, 17, 18; 19:17; 24:1, 11, 16; 34:9, 15, 19; Fig. 77. □1986 *Stenodontus compressus* n.sp. – Chen & Gong, pp. 186–187, Pls. 19:7; 25:2, 5, 8, 11–13, 16; Fig. 77. □1987 *Coelocerodontus bicostatus* van Wamel – An, p. 104, Pl. 1:11. □1987 *Diaphanodus latus* (van Wamel) – Bagnoli *et al.*, p. 155, Pl. 2:11, 12. □1988 *Coelocerodontus cambricus* (Nogami) – Heredia & Bordonaro, p. 190, Pl. 1:2, 3. □1988 *Rotundoconus mendozanus* n.sp. – Heredia & Bordonaro, pp. 194–195, Pls. 3:5; 4:3. □1988 Gen. sp. indet. A – Heredia & Bordonaro, p. 195, Pls. 3:7; 4:4. □1988 *Coelocerodontus* apparatus – Andres, Pl. 9:3–8; Fig. 19. □1988 *Coelocerodontus cambricus* (Nogami) – Lee, B.S. & Lee, H.Y., Pl. 1:27.

Material. – 150 alpha, 210 beta elements.

Occurrence. – Zone I: Gum ; Zone Vb: Brattefors, Stenstorp–Dala?; Zone Vc: Brattefors, Degerhamn, Grönhögen, Haggården–Marieberg, S. Möckleby–Degerhamn, Trolmen; Zone V undiff.: Trolmen.

Description. – Morphotype alpha comprises asymmetrical simple cones with a broad and even recurvature. They may also be slightly deflected laterally. The pointed apex extends far beyond the posterior basal margin. The sclerites are laterally compressed with keeled anterior and posterior edges. The flanks are usually characterized by a costa, each of which extends up to the apex. Position and development of these costae are highly variable. They may be placed either in the anterior, median or posterior section and are in general asymmetrically arranged. They may also fade away in the lower half, leading to highly variable cross-sections. By contrast, the apical part is uniformly rhomboidal in cross-section. In general, the posterior keel points to the less developed flank, which stresses the asymmetry (Pl. 41:7). The basal cavity is very deep and extends over approximately three-fourths of the entire length (Fig. 20). As the completely translucent sclerites are exceedingly thin-walled, the basal rim is often broken. The outer surface may be faintly annulated, the inner one appears smooth.

Morphotype beta consists of coniform elements which are distinctly recurved particularly in their hook-like apical portion. The whole element is laterally compressed. One of the broad flanks is slightly convex, the other one is concave. Both are characterized by longitudinal depressions leading to an undulated cross-section. Anterior and posterior edges are keeled, with the anterior keel slightly deflected towards the concave side. The basal cavity is as large as in morphotype alpha, the basal rim is usually broken. The cross-section passes from subcircular to oval at the apex into elongate asymmetrical at the basis, with a width:height ratio of

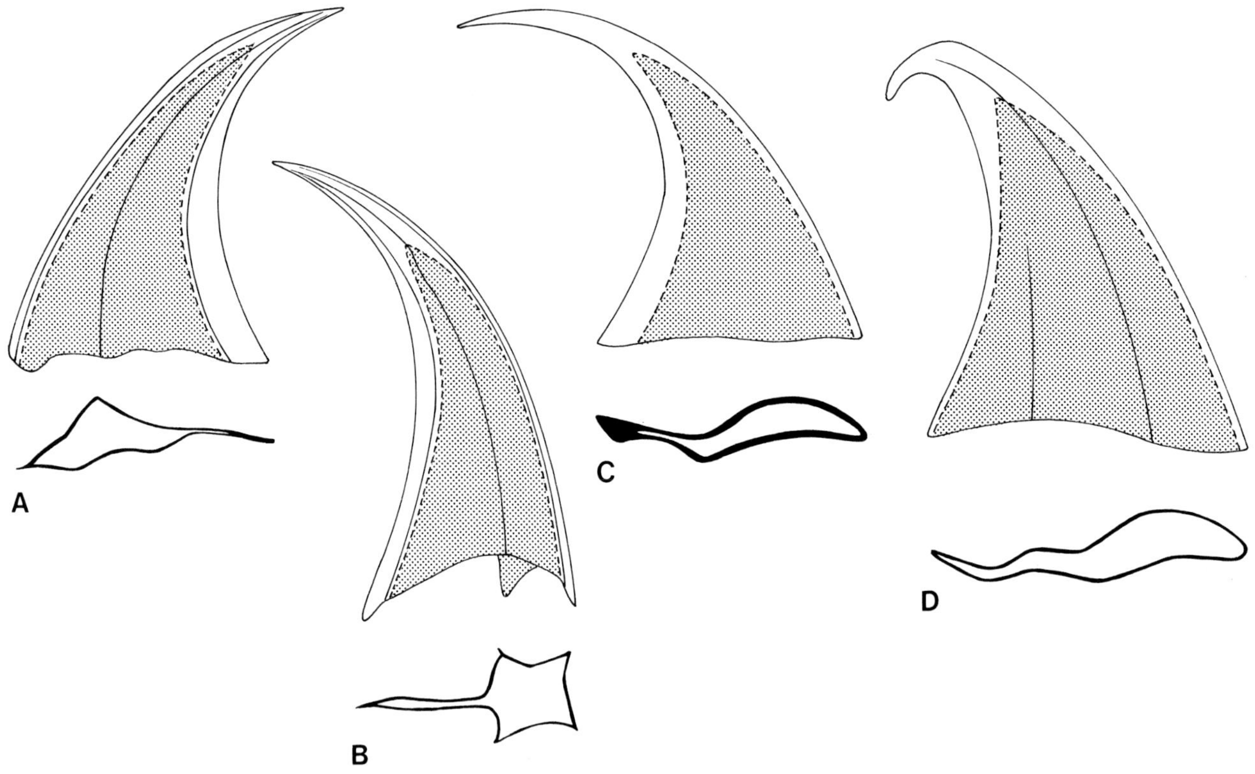

Fig. 20. Coelocerodontus bicostatus van Wamel 1974. □A. Morphotype alpha; ×165. □B. Morphotype alpha; ×140. □C. Morphotype alpha; ×150. □D. Morphotype beta; ×165.

about 5:1. The outer surface may show a gentle annulation, the inner one is smooth.

Size range. – Morphotype alpha: 230–600 μm. Morphotype beta: 370–500 μm.

Comparison. – Morphotype alpha differs from *Distacodus palmeri* in being more laterally compressed and in the asymmetrical arrangement of the costae. It differs from *Proconodontus* in its extremely thin-walled structure, as was already pointed out by Ethington (1959). Further, a seam comparable to the latter genus is absent.

Genus *Cordylodus* Pander 1856

Type species. – *Cordylodus angulatus* Pander 1856.

Cordylodus primitivus Bagnoli, Barnes & Stevens 1987

Pl. 43:4, 5, 8, 9, 11–15; Fig. 21A–F

Synonymy. – □1981 *Cordylodus* sp. – Andres, pp. 23–27 Figs. 11–18. □1985 *Cordylodus* sp. (Andres) – Borovko & Sergeeva, Figs. 16, 17. □1986 *Cordylodus andresi* [nomen nudum] – Viira, *In* Kaljo *et al.*, p. 103, Pl. 2:1–6, 9, 10. □1987 *Cordylodus primitivus* n.sp. – Bagnoli, Barnes & Stevens, p. 154, Pl. 1:1–6. □1987 *Cordylodus andresi* n.sp. – Viira & Sergeeva, pp. 147–148 Pls. 1:1–8; 3:1, 2, 4; Fig. 2:18, 33–36, 42–59; Fig. 4:28. □1988 *Cordylodus andresi* Viira & Sergeeva – Kaljo *et al.*, Figs. 4a, 5a. □non! 1988 *Cordylodus andresi*

n.sp. – Barnes, pp. 410–411, Fig. 13d–f, Fig. 14a (homonym).

Material. – 25 specimens.

Occurrence. – Zone Vc: S. Möckleby–Degerhamn, Trolmen; Zone V undiff.: Degerhamn, Skår, S. Möckleby, Trolmen.

Remarks. – Barnes (1988) described a new species as *Cordylodus andresi*. In our opinion the illustrated specimens are neither *C. primitivus* Bagnoli *et al.* 1987 nor *C. andresi* Viira 1987. Being a junior homonym, *C. andresi* Barnes 1988 should be replaced.

Description. – Compound sclerites with a distinctly recurved main denticle. The anterior side is convex, the posterior one is concave and expanded to a process, which carries up to eight discrete denticles of variable size. Some specimens resemble *Proconodontus* in general outline. Accordingly, the typical process is missing; instead the denticles are developed alongside the steep posterior margin. The basal cavity is large and usually extends beyond the point of flexure of the main denticle. It also runs the length of the posterior process (Fig. 20). Contrary to Andres' material it does not extend into the secondary denticles. The cross-section of the main denticle is oval to lenticular. The basis is somewhat cuneiform and tapers towards the posterior edge. The basal rim is always broken. Outer and inner surface appear smooth.

Size range. – 520–830 μm.

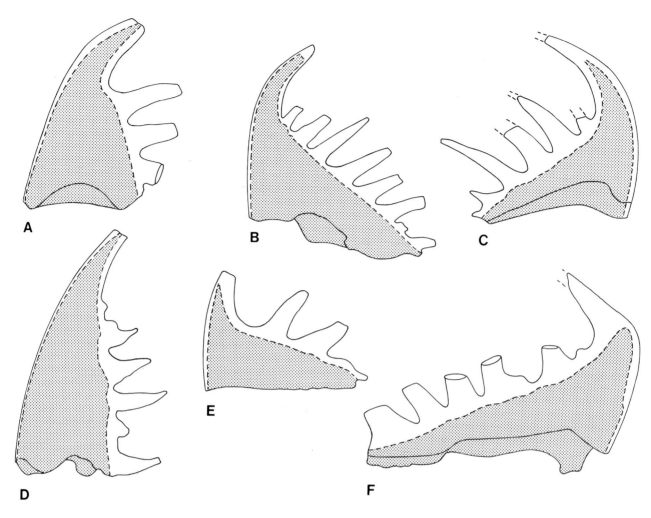

Fig. 21. Cordylodus primitivus Bagnoli, Barnes & Stevens 1987; all ×115.

Comparison. – *Cordylodus primitivus* differs from *C. proavus* in the development of the main denticle which generally seems to be less recurved but more triangular in lateral outline – a character close to *Proconodontus*. The typical 'proavus' element is more reclined combined with more or less parallel sides of the main denticle.

The PRIMITIVUS–OKLAHOMENSIS concept. – Andres (1981) described a large *Cordylodus* association from the *Acerocare* zone of the isle of Öland. It comprises elements which he referred to *C. proavus* s.f. Müller 1959. However, he noted transitions to morphologically very different forms of the genus and therefore refrained from a specific determination. Including part or all his material, Bagnoli *et al.* (1987) erected the new species *C. primitivus,* and Viira (*in* Viira, Sergeeva & Popov, 1987) established the separate three-element *Cordylodus andresi* apparatus which, however, is considered a junior synonym of *C. primitivus.*

　In our opinion there are no defineable criteria on single specimens to distinguish between *C. primitivus* and *C. proavus* in the sense of an apparatus. Further, in the previous record, the variability of the latter is considerable and covers all published 'primitivus' representatives. On the other hand, Andres material as a whole looks different from the 'proavus'/'oklahomensis' associations. In this con-

text, small associations cannot be identified as belonging to one or another of them. It is possible that there are two different species or subspecies, one containing the element *C. oklahomensis* Müller 1959 in the form sense and *C. primitivus* Bagnoli *et al.* (1987) both of which include 'proavus' form elements. It might be possible that the 'oklamomensis' type elements appear at a time when 'primitivus' elements are no longer around. The presence of the 'proavus' type alone would thus be insufficient for precise stratigraphic determinations. Such a differentiation might be due to provincialism. In the present sense, *C. proavus* elements are widely distributed. Based merely on the illustrated evidence it is impossible to sort the previous record, and an according restudy of larger associations is imperative.

Genus *Proconodontus* Miller 1969

Type species. – *Proconodontus muelleri* Miller 1969.

Remarks. – This genus can be recognized by the contrast between the blackish or amber-coloured cone and a covering hyaline layer, which exclusively forms the translucent or white opaque keels. This difference is, however, observable only on uncoated specimens and therefore not on the

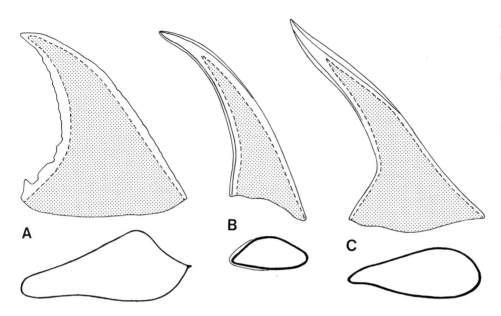

Fig. 22. Outline of two species of *Proconodontus* with basal cavity and cross-section. □A. *P. serratus* (Miller 1969); ×50. □B. *P. muelleri* Miller 1969. Note the seam-like keel at the posterior side of the cross-section; ×100. □C. *P. muelleri* Miller 1980; ×140.

SEM photos. Species are distinguished by the different development of the keels.

According to Miller (1969, 1981) *Proconodontus* represents the most primitive euconodonts and comprises all forms with a very large basal cavity marked by anterior and/or posterior keels. White matter is always lacking. Our material yielded, however, specimens, which clearly show traces of white matter in the keeled translucent seam. The structure resembles very much the one known from *Cordylodus.*

Proconodontus muelleri Miller 1969
Pl. 42:1–16; Fig. 22B, C

Synonymy. – □1969 *Proconodontus muelleri muelleri* n.ssp. – Miller, p. 437, Pl. 66:30–40; Fig. 5H. 9K. □1971 *Proconodontus muelleri muelleri* Miller – Miller & Melby, Pl. 2:18. □1973 *Proconodontus muelleri muelleri* Miller – Lindström, p. 407, Pl. 1:5. □1973 *Proconodontus muelleri* Miller – Müller, p. 42, Pl. 3:4–8, 10. □1978 *Proconodontus muelleri muelleri* s.f. Miller – Tipnis, Chatterton & Ludvigsen, Pl. 1:13, 17 . □1980 *Proconodontus muelleri* Miller – Miller, pp. 29–30, Pl. 1:7. □1981 *Proconodontus muelleri* Miller – Miller *in* Robison, p. W146, Fig. 95.2a, b. □1982 *Proconodontus muelleri* Miller – Fortey, Landing & Skevington, p. 117; Fig. 9K. □1982 *Proconodontus muelleri* Miller – An, pp. 141–142, Pl. 12:8–9, 11, 12. □1983 *Proconodontus muelleri* Miller – An *et al.,* pp. 126–127, Pl. 5:15, 16, 21–24. □1985 *Proconodontus muelleri muelleri* Miller – Nowlan, p. 114, Fig. 5.1. □1985b *Proconodontus muelleri* Miller – Wang, p. 95, Pl. 26:10–13; Fig. 13/1. □1986 *Proconodontus muelleri* Miller – Chen, Zhang & Yu, p. 368, Pl. 2, Fig. 7 ?1986 *Proconodontus muelleri* Miller – Chen & Gong, pp. 159–161, Pl. 19:6; Fig. 60 (part) . □1987 *Proconodontus muelleri* Miller – An, pp. 109–110, Pl. 2:4, 17. □1987 *Proconodontus* aff. *muelleri* Miller – An, p. 110, Pl. 2:19–21. □1987 *Procondoontus posterocostatus* Miller – An, p. 111, Pl. 2:18. □1988 *Proconodontus muelleri* Miller – Andres, pp. 125–128, Pls. 10:1–8; 11:1–8; 12:1–4; Figs. 20–25. □1988 *Proconodontus muelleri* Miller – Lee, B.S. & Lee, H.Y., Pl. 2:27–29.

Material. – 390 specimens.

Occurrence. – Zone Vb: Ödegården; Zone Vc: Grönhögen, Haggården–Marieberg, Ödegården, Trolmen; Zone V undiff.: Degerhamn, Grönhögen, Kalvene, Milltorp, Trolmen.

Description. – Coniform, slightly asymmetrical sclerites that are more or less recurved and slightly deflected to either side. The pointed apex often shows traces of regeneration (Pl. 42:4). Both anterior and posterior sides are characterized by sharp keels of variable extent. They derive from the outermost hyaline layer which covers the whole cone. Usually a keel is developed along the entire posterior side. By contrast, it takes a variable part of the anterior side. The basal cavity is very large and takes more than ⅘ of the entire cone (Fig. 21B). The basal rim has not been preserved on any specimen. The cross-section is circular at the very initial part. Towards the basis it changes into an oval, subelliptical or even tear-shaped outline. The outer surface is smooth to faintly annulated, the inner one is smooth. Occasionally, perforations are visible in that area. Pl. 42:10 shows a specimen with sparsely scattered penetrations following the basalmost annulations. The inner surface is, however, often dotted with circular holes that do not correspond to the outer layer (Pl. 42:13). Left and right forms occur in nearly the same ratio.

Remark. – Part of the material has been with James F. Miller, who confirmed the identification (personal communication, 1990).

Size range. – 360–700 µm.

Proconodontus serratus Miller 1969
Pls. 42:17–21; 43:1–3, 6, 7; Fig. 22A

Synonymy. – □1969 *Proconodontus muelleri serratus* n.ssp. – Miller, p. 438, Pl. 66:41–44. □1973 *Proconodontus serratus* Miller – Müller, p. 44, Pl. 4:1, 2. □1973 *Proconodontus muelleri serratus* Miller – Lindström, p. 409, Pl. 1:6. □1980 *Proconodontus serratus* Miller – Landing, Ludvigsen & von

Bitter, p. 33, Fig. 8I–L. □1980 *Proconodontus serratus* Miller – Miller, p. 31, Pl. 1:13; Fig. 4D. □1982 *Proconodontus muelleri serratus* Miller – Fortey, Landing & Skevington, p. 124, Fig. 9L. □1986 *Proconodontus serratus* Miller – Chen & Gong, pp. 162–163, Pl. 33:1; Fig. 62. □1988 *Proconodontus serratus* Miller – Lee, B.S. & Lee, H.Y., Pl. 2:30, 31.

Material. – 30 specimens.

Occurrence. – Zone Vc: S. Möckleby–Degerhamn, Trolmen; Zone V undiff.: Trolmen.

Description. – Gently and evenly recurved sclerites. In some cases, the basal width corresponds to the length and makes the elements appear rather stout. Anterior and posterior sides are characterized by the typical keel that becomes broader towards the basis. The posterior keel is serrated except for the apical part. Either discrete denticles of different size are developed, or the edge appears gently scalloped; the latter might be regarded as a matter of preservation (Pl. 42:17). The basal rim is always incompletely preserved. The basal cross-section is subtriangular with rounded edges. The other features are well in accord with *Proconodontus muelleri.*

Size range. – 730–1050 μm.

References

Abaimova, G.P. 1978: Pozdnekembrijskie konodonti zentral'noga Kazakhstana. [Late Cambrian conodonts of central Kazakhstan.] *Paleontologicheskij Zhurnal 1978:4,* 77–87.

Abaimova, G.P. 1980: Apparaty kembrijskich konodontov iz Kazakhstana. [Cambrian conodont apparatuses from Kazakhstan.] *Paleontologicheskij Zhurnal 1980:2,* 143–146.

Abaimova, G.P. & Ergaliev, G.Kh. 1976: O nakhodke konodontov v srednem i verkhnem kembrii Malogo Karatau. [On finds of conodonts in the Middle and Upper Cambrian of Malyj Karatau.] *Trudy Instituta Geologii i Geofiziki SO AN SSSR 333,* 390–394.

Abaimova, G.P. & Markov, E.P. 1977: Pervye nakhodki konodontov nizhneordovikskoj zony *Cordylodus proavus* na yuge Sibirskoj Platformy. [The first finds of conodonts in the Lower Ordovician *Cordylodus proavus* Zone in the south of the Siberian Platform.] *In* Sokolov, B.S. & Kanygin, A.V. (eds.): *Problemy Stratigrafii Ordovika i Silura Sibiri. Trudy Instituta Geologii i Geofiziki SO AN SSSR 372,* 86–94. Novosibirsk.

Aldridge, R.J., Briggs, D.E.G., Clarkson, E.N.K. & Smith, M.P. 1986: The affinities of conodonts – new evidence from the Carboniferous of Scotland. *Lethaia 19,* 279–291.

An, T.-X. 1981: Recent progress in Cambrian and Ordovician Conodont Biostratigraphy of China. *Geological Society of America Special Paper 187,* 209–225.

An, T.-X. 1982: [Study on the Cambrian Conodonts from North and Northeast China.] *Science Report of the Institute of Geoscience, Sect. B 3,* 113–159. University of Tsukuba.

An, T.-X. 1987: [*The Lower Paleozoic Conodonts of South China.*] 238 pp. The Publishing House of Beijing University.

An, T.-X., Du, G.-Q. & Gao, Q.-Q. 1985: [*Ordovician Conodonts from Hubei, China.*] 64 pp. Geological Publishing House, Beijing.

An, T.-X. & Yang, C.-S. 1980: Cambro-Ordovician conodonts in North China with special reference to the boundary between the Cambrian and Ordovician systems. *Scientific Papers on Geology for International Exchange 4, Stratigraphy and Paleontology,* 7–14.

An, T.-X., Zhang, F., Xiang, W., Zhang, Y., Xu, W., Zhang, H., Jiang, D., Yang, C., Lin, L., Cui, Z. & Yang, X. 1983: [*The Conodonts of*

North China and the Adjacent Regions.*] 223 pp. Science Press, Beijing.

Andres, D. 1981: Beziehungen zwischen kambrischen Conodonten und Euconodonten. *Berliner geowissenschaftliche Abhandlungen (A) 32,* 19–31.

Andres, D. 1988: Struktur, Apparate und Phylogenie primitiver Conodonten. *Paläontographica A 200(4–6),* 105 pp.

Apollonov, M.K. & Chugaeva, M.N. 1982: Batyrbajsajskij razrez kembriya i ordovika v Malom Karatau (Yuzhnyj Kazakstan). [The Cambrian–Ordovician Batyrbaisai section in Malyj Karatau (South Kazakhstan).] *Izvestiya Akademia Nauk SSSR, Seriya Geologicheskaya 1982(4),* 36–46.

Apollonov, M.K., Chugaeva, M.N. & Dubinina, S.V. 1984: . [*Trilobites and Conodonts from the Batyrbaisai Section (Uppermost Cambrian – Lower Ordovician) in Malyj Karatau Range. (Atlas of the paleontological plates.)*] 48 pp, 32 pls. AN KazSSSR Publishing House, Alma-Ata.

Azmi, R.J. 1983a: Blaini–Krol–Tal: A Late Precambrian – Early Paleozoic Succession of the Lesser Himalaya. *Abstracts for the 5th Indian Geophytological Conference., B.S.I.P.* Lucknow.

Azmi, R.J. 1983b: Microfauna and age of the Lower Tal Phosphorite of Mussoorie Syncline, Garhwal Lesser Himalaya, India. *Himalayan Geology 11,* 273–409 .

Azmi, R.J. & Pancholi, V.P. 1983: Early Cambrian (Tommotian) conodonts and other shelly microfauna from the Upper Krol of Mussoorie Syncline, Garhwal Lesser Himalaya with remarks on the Precambrian–Cambrian boundary. *Himalayan Geology 11,* 360–372.

Azmi, R.J., Yoshi, M.N. & Juyal, K.P. 1981: Discovery of the Cambro-Ordovician conodonts from the Mussoorie Tal Phosphorite: its significance in correlation of the Lesser Himalaya. *Contemporary Geoscientific Researches in Himalaya 1,* 245–250.

Bagnoli, G., Barnes, C.R. & Stevens, R.K. 1987: Lower Ordovician (Tremadocian) conodonts from Broom Point and Green Point, Western Newfoundland. *Estratto dal Bolletino della Società Italiana 25(2),* 145–158.

Barnes, C.R. 1988: The proposed Cambrian–Ordovician global boundary stratotype and point (GSSP) in Western Newfoundland, Canada. *Geological Magazine 125(4),* 381–414.

Bednarczyk, W. 1979: Upper Cambrian to Lower Ordovician conodonts of Leba elevation, NW Poland, and their stratigraphic significance. *Acta Geologica Polonica 29(4),* 409–422.

Bengtson, S. 1976: The structure of some Middle Cambrian conodonts, and the early evolution of conodont structure and function. *Lethaia 9,* 185–206.

Bengtson, S. 1983a: A functional model for the conodont apparatus. *Lethaia 16,* 38.

Bengtson, S. 1983b: The early history of the Conodonta. *Fossils and Strata 15,* 5–19.

Bergström, J. & Gee, D.G. 1985: The Cambrian in Scandinavia. *In* Gee, D.G. & Sturt, B.A. (eds): *The Caledonide Orogen – Scandinavia and Related Areas,* 248–271. John Wiley & Sons Ltd.

Bergström, S.M. 1988: A quantitative analysis of Lower Paleozoic global conodont provincialism. *In* Ziegler, W. (ed.): *Courier Forschungsinstitut Senckenberg 102,* 299.

Bhatt, D.K. 1980: Discovery of conodonts in the Cambrian of Spiti, Tethys Himalaya. *Current Sciences 49(9),* 357–358.

Bischoff, G.C.O. & Prendergast, E.I. 1987: Newly discovered Middle and Late Cambrian fossils from the Wagonga Beds of New South Wales, Australia. *Neues Jahrbuch für Geologie und Paläontologie Abh. 175(1),* 39–64.

Bond, G.C., Nickerson, P.A. & Kominz, M.A. 1984: Breakup of a supercontinent between 625 Ma and 555 Ma: new evidence and implications for continental histories. *Earth and Planetary Science Letters 70,* 325–345.

Borovko, N.G. & Sergeeva, S.P. 1985: Konodonty verkhne-kembrijskikh otlozhenij baltijsko-ladozhskogo glinta. [Conodonts of the Upper Cambrian deposits of the Baltic–Ladoga klint.] *Proceedings of the Academy of Sciences of the Estonian SSR, Geology 34(4),* 125–129.

Brasier, M. 1984: Microfossils and small shelly fossils from the Lower Cambrian *Hyolithes* Limestone at Nuneaton, English Midlands. *Geological Magazine 121(3)*, 229–253.

Bruton, D.L., Koch, L. & Repetski, J.E. 1988: The Naersnes Section, Oslo region, Norway: Trilobite, graptolite and conodont fossils reviewed. *Geological Magazine 125*, 451–455.

Buggisch, W. 1982: Conodonten aus den Ellsworth Mountains (Oberkambrium), Westantarktis. *Zeitschrift der Deutschen Geologischen Gesellschaft 133*, 493–507.

Buggisch, W. & Repetski, J.E. 1987: Uppermost Cambrian(?) and Tremadocian conodonts from Handler Ridge, Robertson Bay Terrane, North Victoria Land, Antarctica. *Geologisches Jahrbuch 66*, 145–185.

Burrett, C.F. & Findlay, R.H. 1984: Cambrian and Ordovician conodonts from the Robertson Bay Group, Antarctica and their tectonic significance. *Nature 307*, 723–726.

Carls, P.C. 1977: Could conodonts be lost and replaced? *Neues Jahrbuch für Geologie und Paläontologie 155(1)*, 18–64.

Chang, W.T., Chu, C.L. & Lin, H.L. 1980: Stages and zones of the Cambrian System in China and its correlation. *Scientific Papers on Geology for International Exchange 4, Stratigraphy and Paleontology*, 1–6.

Chen, J.-Y. & Gong, W.-L. 1986: Conodonts. *In* Chen, J.-Y. (ed.): *Contributions to Dayangcha International Conference on Cambrian/Ordovician Boundary*, 93–223. China Prospect Publishing House, Beijing.

Chen, M.-J., Zhang, J.-H. & Yu, Q. 1986: Cambrian–Ordovician conodonts from the Jiangnan region. *Acta Micropalaeontologica Sinica 3(4)*, 361–372.

Chugaeva, M.N. & Apollonov, M.K. 1982: The Cambrian–Ordovician boundary in the Batyrbaisai Section, Malyi Karatau Range, Kazakhstan, SSSR. *In* Bassett, M.G. & Dean, W.T. (eds.): The Cambrian–Ordovician boundary sections, fossil distributions and correlations. *National Museum of Wales, Geological Series 3*, 77–85.

Clark, D.L. & Miller, J.F. 1969: Early evolution of conodonts. *Geological Society of America, Bulletin 80*, 125–134.

Clark, D.L. & Robison, R.A. 1969: Oldest conodonts in North America. *Paleontology 43(4)*, 1044–1046.

Conway Morris, S. 1976: A new Cambrian lophophorate from the Burgess Shale of British Columbia. *Palaeontology 19(2)*, 199–222.

Conway Morris, S. 1978: Conodont. *McGraw-Hill Yearbook of Science and Technology 1978*, 130–133.

Derby, J.R., Lane, H.R. & Norford, B.S. 1972: Uppermost Cambrian – Basal Ordovician faunal succession in Alberta and correlation with similar sequences in the Western United States. *24th International Geological Conference, Section 7*, 503–512.

Ding, L., Bao, D. & Cao, H. 1987: [Research on the conodonts near the boundary between Cambrian and Ordovician systems in Jiangshan–Yanglingang region, Zheijiang Province.] *Experimental Petroleum Geology 9(1)*, 74–80.

Dong, X.-P. 1984: [Conodont-based Cambrian–Ordovician boundary at Huanghuachang of Yichang, Hubei.] *In: Stratigraphy and Paleontology of System Boundaries in China. Cambrian–Ordovician Boundary 2*, 383–412. Nanjing Institute of Geology and Paleontology. Academia Sinica.

Dong, X.-P. 1986: [Conodont-based Cambrian–Ordovician boundary at Huanghuachang of Yichang, Hubei.] 383–412. *In: Stratigraphy and Palaeontology of Systemic Boundaries in China: Cambrian–Ordovician Boundary (2)*. Anhui Science and Technology Publishing House, Nanjing.

Dong, X.-P. 1987: [Late Cambrian and Early Ordovician conodonts from Chuxian, Anhui.] *Papers of Graduate Students of Nanjing Institute of Geology and Paleontology*, 135–184. Academia Sinica.

[Dong, X.-P. 1988: Upper Middle and Lower Cambrian trilobite and conodont biostratigraphy in Huayuan, Hunan. Ph.D. Thesis, University of Beijing.]

Dong, X.-P. 1990: [A potential candidate for the Middle–Upper Cambrian boundary stratotype – An introduction to the Paibi section in Huayuan, Hunan.] *Acta Geologica Sinica 1990(1)*, 62–79.

Druce, E.C. & Jones, P.J. 1968: Stratigraphical significance of conodonts in the Upper Cambrian and Lower Ordovician sequence of the Boulia Region, Western Queensland. *The Australian Journal of Science 31*, 88.

Druce, E.C. & Jones, P.J. 1971: Cambro-Ordovician conodonts from the Burke River structural belt Queensland. *Bureau of Mineral Resources, Geology and Geophysics, Bulletin 110*, 1–159.

Druce, E.C., Shergold, J.H. & Radtke, B.M. 1982: A reassessment of the Cambrian–Ordovician boundary section at Black Mountain, Western Queensland, Australia. *In* Bassett, M.G. & Dean, W.T. (eds.): The Cambrian–Ordovician boundary: sections, fossil distributions and correlations. *National Museum of Wales Geological Series 3*, 193–209.

Dubinina, S.V. 1982: Konodontovyje assotsiatii pogranitchnykh otlozhenij kembrija i ordovika Malogo Karatau (yuzhnyj Kazakhstan). [Conodont assemblages of the Cambrian–Ordovician boundary beds of Malyj Karatau (southern Kazakhstan).] *Izvestiya Akademii Nauk SSSR, Seriya Geologicheskaya 4*, 47–54.

Dutro, Jr. J.T., Palmer, R.A., Repetski, J.E. & Brosgé, W.P. 1984: Middle Cambrian fossils from the Doonerak anticlinorium, central Brook Range, Alaska. *Journal of Paleontology 58(6)*, 1364–1372.

Dzik, J. 1986: Chordate affinities of the conodonts. *In* Hoffman, A, & Nitecki, M. H. (eds.) 1986: *Problematic Fossil Taxa*, 241–254. Oxford University Press, New York, Clarendon Press, Oxford.

Eichenberg, W. 1930: Conodonten aus dem Culm des Harzes. *Paläontologische Zeitschrift 12*, 177–182.

Eisenack, A. 1958: Mikrofossilien aus dem Ordovizium des Baltikums. *Senckenbergiana lethaea 39(5/6)*, 389–405.

Erdtmann, B. & Miller, J.F. 1981: Eustatic control of lithofacies and biofacies changes near the base of the Tremadocian. *In* Taylor, M.E. (ed.): Short papers for the 2nd International Symposium on the Cambrian System. *US Geological Survey Open-File-Report 81-743*, 78–81.

Ethington, R. 1959: Conodonts from the Ordovician Galena Formation. *Journal of Paleontology 33*, 257–292.

Fåhraeus, L.E. & Nowlan, G.S. 1978: Franconian (Late Cambrian) to Early Champlainian (Middle Ordovician) conodonts from the Cow Head Group, Western Newfoundland. *Journal of Paleontology 52*, 444–471.

Fortey, R.A., Landing, E. & Skevington, D. 1982: Cambrian–Ordovician boundary sections in the Cow Head Group, Western Newfoundland. *In* Bassett, M.G. & Dean, W.T. (eds.): The Cambrian–Ordovician boundary: sections, fossil distributions, and correlations. *National Museum of Wales Geological Series 3*, 95–129.

Grant, R.E. 1965: Faunas and stratigraphy of the Snowy Range Formation (Upper Cambrian) in southwestern Montana and northwestern Wyoming. *Geological Society of America, Memoir 96*, 143–164.

Gross, W. 1954: Zur Conodonten-Frage. *Senckenbergiana lethaea 35*, 73–85.

Hass, W.H. 1941: Morphology of conodonts. *Journal of Paleontology 15(1)*, 71–81.

Heinsalu, H., Viira, V., Mens, K. & Puura, I. 1987: Kembrijsko-ordovikskie pogranichnye otlozheniya razreza Yulgaze, severnaya Ehstonia. [The section of the Cambrian–Ordovician boundary beds in Ülgase, Northern Estonia.] *Proceedings of the Academy of Science of the Estonian SSSR, Geology 36(4)*, 153–165.

Henningsmoen, G. 1957: The trilobite family Olenidae. *Skrifter utgitt av det Norske Videnskaps-Akademi i Oslo 1. Mat.-Naturv. Klasse. 1.* 362 pp.

Henningsmoen, G. 1958: The Upper Cambrian Faunas of Norway. With descriptions of non-Olenid invertebrate fossils. *Norsk Geologisk Tidsskrift 38(2)*, 179–196.

Heredia, S. & Bordonaro, O. 1988: Conodontes de la formacion La Cruz (Cambrico Superior), San Isidro, provincia de Mendoza, R. Argentina. *IV Congressisto Argentino de Paleontologia y Biostratigrafia, Actas 3*, 189–202.

Hintze, L.F., Taylor, M.E. & Miller, J.F. 1988: Upper Cambrian–Lower Ordovician Notch Peak Formation in Western Utah. *US Geological Survey, Professional Paper 1393*, 1–30.

Hinz, I. 1987: The Lower Cambrian microfauna of Comley and Rushton, Shropshire/England. *Palaeontolographica A 198*, 41–100.

Jeppsson, L. 1979: Conodont element function. *Lethaia 12(2)*, 153–171.

Jiang, W., Zhang, F., Zhou, X., Xiong, J., Dai, J. & Zhong, D. 1986: [Conodonts – Palaeontology.] *Southwestern Petroleum Institute*, 1–287.

Jones, P.J. 1961: Discovery of conodonts in the Upper Cambrian of Queensland. *Australian Journal of Science 24*, 143–144.

Jones, P.J. 1971: Lower Ordovician conodonts from the Bonaparte Gulf Basin and the Daly River Basin, Northwestern Australia. *Bureau of Mineral Resources, Geology and Geophysics, Bulletin 117*, 80.

Jones, P.J., Shergold, J.H. & Druce, E.C. 1971: Late Cambrian and Early Ordovician Stages in Western Queensland. *Journal of the Geological Society of Australia 18(1)*, 1–32.

Kaljo, D., Borovko, N., Heinsalu, H., Hazanovits, K., Mens, K., Popov, L., Sergejeva, S., Sobolevskaja, R. & Viira, V. 1986: Kambriumi ja ordoviitsiumi piir Balti-Laadoga klindi piirkonnas (Põhja-Eesti ja Leningradi oblast, NSV Liit. [The Cambrian–Ordovician boundary in the Baltic Ladoga Clint Area (North Estonia and Leningrad region, USSR).] *Proceedings of the Academy of Science of the Estonian SSSR. Geology 35(3)*, 97–108.

Kaljo, D., Heinsalu, H., Mens, K., Puura, I. & Viira, V. 1988: Cambrian–Ordovician boundary beds at Tõnismägi, Tallinn, North Estonia. *Geological Magazine 125(4)*, 457–463.

Kurtz, V.E. 1976: Biostratigraphy of the Cambrian and lowest Ordovician Bighorn Mountains and associated uplifts in Wyoming and Montana. *Brigham Young University Geological Studies 23(2)*, 215–227.

Landing, E. 1974: Early and Middle Cambrian conodonts from the Taconic Allochton, Eastern New York. *Journal of Paleontology 48*, 1241–1248.

Landing, E. 1976: Early Ordovician (Arenigian) conodont and graptolite biostratigraphy of the Taconic Allochton, eastern New York. *Journal of Paleontology 50(4)*, 614–646.

Landing, E. 1977: 'Prooneotodus' tenuis (Müller 1959) apparatuses from the Taconic Allochton, Eastern New York: construction, taphonomy and the protoconodont 'supertooth' model. *Journal of Paleontology 51(6)*, 1072–1084.

Landing, E. 1979: Conodonts and biostratigraphy of the Hoyt Limestone (Late Cambrian, Trempealeauan), Eastern New York. *Journal of Paleontology 53(4)*, 1023–1029.

Landing, E. 1980: Late Cambrian – Early Ordovician macrofaunas and phosphatic microfaunas, St. John Group, New Brunswick. *Journal of Paleontology 54(4)*, 752–761.

Landing, E. 1983: Highgate Gorge: Upper Cambrian and Lower Ordovician continental slope deposition and biostratigraphy, Northwestern Vermont. *Journal of Paleontology 57(6)*, 1149–1188.

Landing, E., Ludvigsen, R. & von Bitter, P.H. 1980: Upper Cambrian to Lower Ordovician conodont biostratigraphy and biofacies, Rabbitkettle Formation, District of Mackenzie. *Life Science Contributions, Royal Ontario Museum 126*, 1–42.

Landing, E., Taylor, M.E. & Erdtmann, B. 1978: Correlation of the Cambrian–Ordovician boundary between the Acado-Baltic and North American faunal provinces. *Geology 6*, 75–78.

Lee, B.-S. & Lee, H.-Y. 1988: Upper Cambrian conodonts from the Hwajeol Formation in the Southern limb of the Baegunsan Syncline, Eastern Yeongweol and Samcheog areas, Kangweon-Do, Korea. *Journal of the Geological Society of Korea 24*, 356–375.

Lee, H.-Y. 1975: Conodonts from the Upper Cambrian Formation, Kangweon-Do, South Korea and its stratigraphical significance. *The Graduate School, Yonsei University, Seoul, Korea, 12*, 71–89.

Lee, H.-Y. 1980: Lower Paleozoic conodonts in South Korea. *Geology and Palaeontology of Southeast Asia 21*, 1–9.

Lindström, M. 1964: *Conodonts*. 196 pp. Elsevier Publishing Company, Amsterdam.

Lindström, M. 1973: *In* Ziegler, W. (ed.): *Catalogue of conodonts IV*. E. Schweitzerbart'sche Verlagsbuchhandlung, Stuttgart, 407, 409, Pl. 1.

Lindström, M. 1974: The conodont apparatus as a food-gathering mechanism. *Palaeontology 17(4)*, 729–744.

Lu, Y.-H. & Mu, E.-Z. 1980: Boundaries of the Ordovician system in China. *Scientific Papers on Geology for International Exchange 4, Stratigraphy and Paleontology*, 15–25.

MacKinnon, D.I. 1976: Middle Cambrian conodonts from New Zealand (Note). *New Zealand Journal of Geology and Geophysics 19(2)*, 294.

Meshkova, N.P. 1969: K voprosu o paleontologicheskoj kharakteristike nizhnego kembriya Sibirskoj platformy. [On the question of the paleontologic characteristics of the Lower Cambrian of the Siberian Platform.] *In* Zhuravleva, I.T. (ed.): *Biostratigraphy and paleontology of the Lower Cambrian of Siberia and the Far East*, 158–174. Institute of Geology and Geophysics of the Academy of Science SSSR, Siberian Section.

Miller, J.F. 1969: Conodont fauna of the Notch Peak Limestone (Cambro-Ordovician) House Range, Utah. *Journal of Paleontology 43*, 413–439.

[Miller, J.F. 1970: Conodont evolution and biostratigraphy of the Upper Cambrian and Lowest Ordovician. Ph.D. Thesis, University of Wisconsin.]

Miller, J.F. 1976: An evolutionary transition between paraconodonts and conodontophorids from the Wilberns Formation (Upper Franconian) of Central Texas. *Geological Society of America, Abstracts with Programs 8*, 498.

Miller, J.F. 1978: Upper Cambrian and Lowest Ordovician conodont faunas of the House Range, Utah. *In* Miller, J.F. (ed.): Upper Cambrian to Middle Ordovician Conodont Fauna of Western Utah. *Southeast Missouri State University Geoscientific Series 5*, 1–33.

Miller, J.F. 1980: Taxonomic revisions of some Upper Cambrian and Lower Ordovician conodonts with comments on their evolution. *Paleontological Contributions 99*, 39. University of Kansas.

Miller, J.F. 1981: Paleozoogeography and biostratigraphy of Upper Cambrian and Tremadocian conodonts. *US Geological Survey Open-File-Report 81-743*, 134–137.

Miller, J.F. 1984: Cambrian and earliest Ordovician conodont evolution, biofacies and provincialism. *Geological Society of America Special Paper 196*, 43–68.

Miller, J.F. 1987: Upper Cambrian and basal Ordovician conodont faunas of the southwest flank of the Llano Uplift, Texas. *21st. Annual Meeting, South Central Section, Geological Society of America, Waco, Texas*, 1–22, 95–100.

Miller, J.F. 1988a: Conodonts as biostratigraphic tools for redefinition and correlation of the Cambrian–Ordovician boundary. *Geological Magazine 125(4)*, 349–362.

Miller, J.F. 1988b: The oldest euconodont extinction crisis (base of the *Cordylodus proavus* Zone) and its relationship to the event stratigraphy. *In* Ziegler, W. (ed.): *Courier Forschungsinstitut Senckenberg 102*, 249.

Miller, J.F. & Kurtz, V.E. 1979: Reassignment of the Dolomite Point Formation of East Greenland from the Middle Cambrian to the Lower Ordovician based on conodonts. *Geological Society of America, Abstracts with Programs 11*, 480.

Miller, J.F. & Melby, J.H. 1971: Trempealeauan Conodonts. *In* Clark, D.L. (ed.): *Conodonts and biostratigraphy of the Wisconsin Paleozoic. Wisconsin Geological Survey, Circular 19*, 4–9.

Miller, J.F., Robison, R.A. & Clark, D.L. 1974: Correlation of Tremadocian conodont and trilobite faunas, Europe and North America. *Geological Society of America, Abstracts with Programs 6*, 1048–1049.

Miller, J.F. & Rushton, A.W.A. 1973: Natural conodont assemblages from the Upper Cambrian of Warwickshire, Great Britain. *Geological Society of America, Abstracts with Programs 5(4)*, 337–338.

Miller, J.F., Taylor, M.E., Stitt, J.H., Ethington, R.L., Hintze, L.F. & Taylor, J.F. 1982: Potential Cambrian–Ordovician boundary stratotype sections in the Western United States. *In* Bassett, M.G. & Dean, W.T. (eds.): The Cambro-Ordovician boundary: sections, fossil distributions and correlations. *National Museum of Wales Geological Series 3*, 155–180.

Miller, R.H. & Paden, E.A. 1976: Upper Cambrian stratigraphy and conodonts from Eastern California. *Journal of Paleontology 50(4)*, 590–597.

Miller, R.H., Sundberg, F.A., Harma, R.H. & Wright, J. 1981: Late Cambrian stratigraphy and conodonts of southern Nevada. *Alcheringa 5(3)*, 183–197.

Missarzhevsky, V.V. 1973: Konodontoobraznye organizmy iz pogranichnykh sloev kembriya i dokembriya Sibirskoj Platformy i Kazakhstana. [Conodont-shaped organisms from Precambrian–Cambrian boundary beds of the Siberian Platform and Kazakhstan.] *In* Zhuravleva, I.T. (ed.): Problemy Paleontologii i Biostratigrafii nizhnego kembriya Sibiri i Dal'nego Vostoka. *Trudy Instituta Geologii i Geofiziki SO AN SSSR 49*, 53–59.

Missarzhevskij, V.V. & Mambetov, A.M. 1981: Stratigrafiya i fauna pogranichnykh sloev kembriya i dokembriya Malogo Karatau. [Stratigraphy and fauna of the Precambrian–Cambrian boundary beds in Malyj Karatau.] *Trudy Geologicheskogo Instituta AN SSSR 326*, 1–92.

Müller, K.J. 1956: Taxonomy, nomenclature, orientation, and stratigraphic evaluation of conodonts. *Journal of Paleontology 30(6)*, 1324–1340.

Müller, K.J. 1959: Kambrische Conodonten. *Zeitschrift der Deutschen Geologischen Gesellschaft 111*, 434–485.

Müller, K.J. 1960: Wert und Grenzen der Conodontenstratigraphie. *Geologische Rundschau 49*, 83–92, 2 Pls.

Müller, K.J. 1962: Supplement to systematics of conodonts. *In* Moore, R.C. (ed.): *Treatise on Invertebrate Paleontology W (Miscellanea)*, W246–W249. Geological Society of America and University of Kansas Press.

Müller, K.J. 1964: Conodonten aus dem unteren Ordovizium von Südkorea. *Neues Jahrbuch für Geologie und Paläontologie, Abhandlung 119(1)*, 93–102.

Müller, K.J. 1971: Cambrian Conodont Faunas. *Geological Society of America Memoir 127*, 5–20.

Müller, K.J. 1973: Late Cambrian and Early Ordovician conodonts from Northern Iran. *Geological Survey of Iran, Report 30*, 5–76.

Müller, K.J. 1979: Phosphatocopine ostracodes with preserved appendages from the Upper Cambrian of Sweden. *Lethaia 12*, 1–27.

Müller, K.J. 1985: Exceptional preservation in calcareous nodules. *Philosophical Transactions of the Royal Society of London 11*, 67–73.

Müller, K.J. & Andres, D. 1976: Eine Conodontengruppe von *Prooneotodus tenuis* (Müller 1959) in natürlichem Zusammenhang aus dem oberen Kambrium von Schweden. *Paläontologische Zeitschrift 50(3/4)*, 193–200.

Müller, K.J. & Nogami, Y. 1971: Über den Feinbau der Conodonten. *Memoirs of the Faculty of Science, Series of Geology and Mineralogy 38(1)*. 87 pp. University of Kyoto.

Müller, K.J. & Nogami, Y. 1972a: Growth and function of conodonts. *Abstract for the 24th International Geological Congress 1972, Section 7*, 20–27.

Müller, K.J. & Nogami, Y. 1972b: Entöken und Bohrspuren bei den Conodontophorida. *Paläontologische Zeitschrift 46*, 68–86.

Müller, K.J. & Walossek, D. 1988: External morphology and larval development of the Upper Cambrian maxillopod *Bredocaris admirabilis*. *Fossils and Strata 23*. 70 pp.

Nogami, Y. 1966: Kambrische Conodonten von China, Teil 1, Conodonten aus den oberkambrischen Kushan-Schichten. *Memoirs of the College of Science, Series B 32(4)*, 351–366. University of Kyoto.

Nogami, Y. 1967: Kambrische Conodonten von China, Teil 2, Conodonten aus den hoch oberkambrischen Yencho-Schichten. *Memoirs of the College of Science Series B 33*, 211–218. University of Kyoto.

Nowlan, G.S. 1985: Late Cambrian and Early Ordovician conodonts from the Franklinian Miogeosyncline, Canadian Arctic Islands. *Journal of Paleontology 59(1)*, 96–122.

Orchard, M.J. 1985: History of conodont studies. Conodonts of the Cambrian and Ordovician Systems from the British Isles. *In* Higgins, A.C. & Austin, R.L. (eds.): *A stratigraphical Index of Conodonts, Chapter 2, Section 2.2*, 32–67. Ellis Horwood Ltd., Chichester.

Orndorff, R.C. 1988: Latest Cambrian and Earliest Ordovician conodonts from the Conococheague and Stonehenge Limestones of Northwestern Virginia. *US Geological Survey Bulletin 1837, Chapter A*, A1–A18.

Özgül, N. & Gedik, I. 1973: New data on the stratigraphy and the conodont faunas of Caltepe Limestone and Seydisehir Formation, Lower Paleozoic of Central Taurus Range. *Turkish Geological Association Bulletin 16*, 39–52.

Pander, C.H. 1856: Monographie der fossilen Fische des Silurischen Systems der Russisch-Baltischen Gouvernements. *Königliche Akademie Wissenschaften, St. Petersburg*, 1–91.

Poulsen, V. 1966: Early Cambrian distacodontid conodonts from Bornholm. *Biologiske Meddelelser, Det Kongelige Danske Videnskabernes Selskab 23(15)*, 1–9.

Repetski, J.E. & Szaniawski, H. 1981: Paleobiologic interpretation of Cambrian and Earliest Ordovician conodont natural assemblages. *In* Taylor, M.E. (ed.): Short Papers for the 2nd International Symposium on the Cambrian System. *US Geological Survey. Open-File-Report 81-743*, 169–172.

Robison, R.A. (ed.) 1981: *Treatise on Invertebrate Paleontology, Part W (Supplement 2, Conodonta)*. Geological Society of America and University of Kansas Press, W1–W202.

Schrank, E. 1973: Trilobiten aus Geschieben der oberkambrischen Stufen 3–5. *Paläontologische Abhandlungen IV*, 805–857.

Scotese, C., Bambach, R.K., Colleen, B., Voo, R. van der & Ziegler, A. 1979: Paleozoic base maps. *Journal of Geology 87*, 217–277.

Serpagli, E. 1974: Lower Ordovician conodonts from Precordilleran Argentina (Province of San Juan). *Bolletino della Società Paleontologica Italiana 13 (1–2)*, 17–98.

Smith, M.P., Briggs, D.E.G. & Aldridge, R.J. 1987: A conodont animal from the Lower Silurian of Wisconsin, USA, and the apparatus of panderodontid conodonts. *In* Aldridge, R.J. (ed.): *Paleobiology of Conodonts*, 91–104.

Stouge, S. & Bagnoli, G. 1988: Early Ordovician conodonts from Cow Head Peninsula, western Newfoundland. *Palaeontographica Italica 75*, 89–179.

Sweet, W.C. 1985: Conodonts: Those fascinating little whatzits. *Journal of Paleontology 59(3)*, 485–494.

Szaniawski, H. 1971: New species of Upper Cambrian conodonts from Poland. *Acta Palaeontologica Polonica 16(4)*, 401–413.

Szaniawski, H. 1980a: Conodonts from the Tremadocian chalcedony beds, Holy Cross Mountains, Poland. *Acta Palaeontologica Polonica 25(1)*, 101–121.

Szaniawski, H. 1980b: Fused clusters of paraconodonts. *In* Schönlaub, H.P. (ed.): *Abstract for the 2nd European Conodont Symposium (ECOS II)*. *Abhandlungen der Geologischen Landesanstalt Wien 35*, 211.

Szaniawski, H. 1982: Chaetognath gasping spines recognized among Cambrian protoconodonts. *Journal of Paleontology 56(3)*, 806–810.

Szaniawski, H. 1983: Structure of protoconodont elements. *Fossils and Strata 15*, 21–27.

Szaniawski, H. 1987: Preliminary structural comparisons of protoconodont, paraconodont and euconodont elements. *In* Aldridge, R.J. (ed.): *Paleobiology of Conodonts*, 35–47. Ellis Horwood, Chichester.

Szaniawski, H. & Bengtson, S. 1988: Formation of the first euconodont elements. *In* Ziegler, W. (ed.): *Courier Forschungsinstitut Senckenberg 102*, 256–257.

Taylor, M.E & Landing, E. 1982: Biostratigraphy of the Cambrian–Ordovician transition in the Bear River Range, Utah and Idaho, Western United States. *In* Basset, M.G. & Dean, W.T. (eds.): The Cambro-Ordovician boundary: sections, fossil distributions and correlations. *National Museum of Wales Geological Series 3*, 181–191.

Taylor, M.E., Repetski, J.E. & Sprinkle, J. 1981: Paleontology and biostratigraphy of the Whipple Cave Formation and Lower House Limestone, Sawmill Canyon, Egan Range, Nevada. *2nd International Symposium on the Cambrian System, Guidebook for Field Trip 1*, 73–77. Denver, Colorado.

Tipnis, R.S. & Chatterton, B.D.E. 1979: An occurrence of the apparatus of 'Prooneotodus' (Conodontophorida) from the Road River Formation, Northwest Territories. *In:* Current Research Part B. *Geological Society of Canada, Paper 79-1B*, 259–262.

Tipnis, R.S., Chatterton, B.D.E. & Ludvigsen, R. 1978: Ordovician conodont biostratigraphy of the southern district of Mackenzie, Canada. *In* Stelck, C.R. & Chatterton, B.D.E. (eds.): Western and arctic Canadian biostratigraphy. *Geological Association of Canada, Special Paper 18*, 39–91.

Viira, V., Sergeeva, S. & Popov, L. 1987: Samye rannie predstaviteli roda *Cordylodus* (Conodonta) iz severnoj Ehstonii i Leningradskoj Oblasti. [Earliest representatives of the genus *Cordylodus* (Conodonta) from Cambro-Ordovician boundary beds of North Estonia and Leningrad Region.] *Proceedings of the Academy of Sciences of the Estonian SSR, Geology 36(4)*, 145–153.   ·

Wamel, V.A. van 1974: Conodont biostratigraphy of the Upper Cambrian and Lower Ordovician of northwestern Öland, southeastern Sweden. *Utrecht Micropaleontological Bulletin 10*, 1–126.

Wang, C.-Y. & Li, D.J. 1986: Conodonts from the Ertaogou Formation in Central Jilin Province. *Acta Micropalaeontologica Sinica 3(4)*, 421–428.

Wang, Z.-H. 1985a: Late Cambrian and Early Ordovician conodonts from North and Northeast China with comments on the Cambro-Ordovician boundary. *Stratigraphy and Paleontology of Systemic Boundaries in China, Cambro-Ordovician Boundary*, 195–238.

Wang, Z.-H. 1985b: Conodonts. *In* Chen, J.-Y., Qian, Y.-Y., Lin, Y.-K., Zhang, J.-M., Wang, Z.-H., Yin, L.-M. & Erdtmann, B.: [*Study on Cambrian–Ordovician Boundary Strata and its biota in Dayangcha,*

Hunjiang, Jilin, China], 83–101. China Prospect Publishing House.

Westergård, A.H. 1922: Sveriges Olenidskiffer. *Sveriges Geologiska Undersökning C 18*, 205 pp.

Westergård, A.H. 1953: Two problematic fossils from the Cambrian in Sweden. *Geologiska Föreningens i Stockholm Förhandlingar 75(4)*, 465–468.

Wiman, C. 1893: Über das Silurgebiet des Bottnischen Meeres. *Bulletin of the Geological Institution of the University of Uppsala 1*, 65–75.

Wiman, C. 1903: Studien über das nordbaltische Silurgebiet. I. Olenellussandstein, Obolussandstein und Ceratopygeschiefer. *Bulletin of the Geological Institution of the University of Uppsala 6*, 12–76.

Wright, T.O., Ross, R.J. & Repetski, J.E. 1984: Newly discovered youngest Cambrian or oldest Ordovician fossils from the Robertson Bay Terrane (formerly Precambrian), Northern Victoria Land, Antarctica. *Geology 12*, 301–305.

Zhao, D.-A. 1986: [Conodonts.] *In*: [Cambrian–Ordovician boundary and its interval biotas, southern Jilin, northeast China.] *Journal of the Changchun College of Geology, Special Issue Stratigraphy and Paleontology, Changchun*, 87–110.

Appendix: Locality list

The coordinates are given as six-digit numbers representing degrees (two first digits), minutes (two middle digits) and seconds (two last digits) of latitude and longitude, respectively.

Backeborg, Kinnekulle	N583216	E131958	Zone I
Brattefors, Kinnekulle	N583250	E132446	Zones II, Vb, c, undiff.
Degerhamn, Isle of Öland	N562103	E162451	Zones I–IV, Va?, c, undiff.
Djupadalen, Falbygden	N581021	E133838	Zone Vc
Ekebacka, Kinnekulle	N583201	E132014	Zone II
Ekeberget, Falbygden	N581105	E132504	Zone Vc, undiff.
Ekedalen, Falbygden	N581241	E135035	Zones III, Vc, undiff.
Eriksöre, Isle of Öland	N563631	E162846	Zone Vb
Gössäter, Kinnekulle	N583637	E132642	Zones I, II
Grönhögen, Isle of Öland	N561623	E162417	Zones II–Vb, c, undiff.
Gudhem, Falbygden	N581457	E133247	Zone I
Gum, Kinnekulle	N583155	E132040	Zones I, Vb, c, undiff.
Haggården–Marieberg, Kinnekulle	N583355	E132601	Zones II, Vb, c, undiff.
Kakeled, N of Blomberg, Kinnekulle	N583305	E132011	Zones I, V undiff.
Karlsfors, Falbygden	N582938	E134451	Zones II, III, V undiff.
Karlsro, Falbygden	N582241	E134906	Zone Vc
Kleva, Falbygden	N581237	E132637	Zones I, III, IV
Klippan, N of Blomberg, Kinnekulle	N583306	E132001	Zones I, II
Ledsgården, Falbygden	N581120	E132503	Zone II
Milltorp, Falbygden	N580642	E134313	Zones I, Vb, c
Mörbylilla, Isle of Öland	N561834	E162450	Zone V undiff.
Mossebo, Hunneberg	N582129	E123037	Zone III
Nästegården, Falbygden	N581056	E132552	Zones I, III, Vc
Nya Dala, Falbygden	N581535	E134523	Zone Vc, undiff.
Nygård, Hunneberg	N581939	E122341	Zones III, IV
Ödbogården, Kinnekulle	N583328	E132525	Zones IV, Va?, b, c, undiff.
Ödegården, E of Varv, Kinnekulle	N581137	E135004	Zone Va, b, c, undiff.
Österplana, Kinnekulle	N583428	E132628	Zones I–III, V
Pusabäcken, Kinnekulle	N583408	E132015	Zone II
Ranstadsverket, Falbygden	N581834	E134340	Zone Vb, c, undiff.
Rörsberga, Falbygden	N581350	E133703	Zone Vb, c, undiff.
Sätra, S of Österplana, Kinnekulle	N583416	E132552	Zone I
Sandtorp, Kinnekulle	N583245	E132332	Zones II, Vc?, undiff.
Skar, Falbygden	N581135	E132538	Zones II–Vb, c, undiff.
Smedsgården–Stutagården, Falbygden	N581334	E134949	Zone Vb, c, undiff.
Södra Möckleby, Isle of Öland	N562108	E162546	Zones I–III, Vb, c, undiff.
Stenåsen, Falbygden	N581524	E134630	Zone Vc, undiff.
Stenstorp–Dala, Falbygden	N581549	E134531	Zone Va?, b, c, undiff.
Stora Backor, Falbygden	N581313	E132814	Zone IV?
Stora Stolan, Falbygden	N583124	E134815	Zones I–III, Vc, undiff.
Stubbegården, Kinnekulle	N583209	E132158	Zones I, II, Vc
Tomten, Falbygden	N581328	E131649	Zone Va, b, c, undiff.
Toreborg, Kinnekulle	N583541	E133646	Zone II
Trolmen, Kinnekulle	N583541	E132118	Zones I–III, Vb, c, undiff.
Uddagården, Falbygden	N581110	E133829	Zone Vc, undiff.

Erratic boulders (North Germany)

Alt-Hüttendorf, Mark Brandenburg	Zone II
Beggerow, Pommern	Zone III
Berlin–Spandau	Zone II
Berlin–Wilhemshorst	Zone III
Blankenfelde, Berlin	Zone II
Dwasiden–Hülsenkrug, Isle of Rügen	Zone V
Isle of Fehmarn	Zone II
Hiddensee	Zone V
Kuhbier, Mark Brandenburg	Zone V
Mucran, Isle of Rügen	Zone II
Oderberg, Mark Brandenburg	Zone I
Mark Brandenburg	Zone V (see Müller 1959, p. 478)
Sassnitz, Isle of Rügen	Zone II
Sellin, Isle of Rügen	Zone V
Stoltera, Mecklenburg	Zone V
Weissenhaus, Ostsee	Zone II

Plates

Plate 1

1–5, 7–9, 12–14, 22: *Phakelodus elongatus* (An 1983), p. 32.

☐ 1. UB 933 (sample 5948): Stenstorp–Dala, zone Vb. Lateral view of isolated sclerite.

☐ 2–4, 13. UB 934 (sample 6763): Gum, zone I. Almost complete apparatus with ca 26 sclerites.
2–4. Different side views showing the variable size of individual sclerites within an apparatus.
13. View from above.

☐ 5. UB 935 (sample 6771): Gum, zone I.
Oblique lateral view of fragmentary cluster.

☐ 7, 12. UB 936 (sample 6760): Gum, zone I.
7. Oblique side view of slightly deformed cluster, shows adhesion of elements by secondary phosphatic coating.
12. View onto the basis with tear-shaped cross-sections.

☐ 8. UB 937 (sample 6753): Gum, zone I.
Oblique lateral view on a partly opened cluster.

☐ 14, 22. UB 938 (sample 6398): Haggården–Marieberg zone II.
22. Lateral view of comparatively broad sclerite with coarsely wrinkled outer layer.
14. Detail of the surface.

6, 10, 11, 15–21, 23: *Phakelodus tenuis* (Müller 1959), p. 33.

☐ 6. UB 939 (sample 6369): Brattefors, zone V undiff. Side view of small cluster.

☐ 10, 15. UB 940 (sample 5965): Trolmen, zone V undiff. Lateral view and cross-section of fragmentary cluster.

☐ 11, 16. UB 941 (sample 5965): Trolmen, zone V undiff.
16. Oblique view from the posterior.
11. Detail from the fractured tip, exposing the comparatively thick wall and large basal opening.

☐ 17–19. UB 942 (sample 926): Grönhögen, zone Vc.
17, 18. Lateral and posterior views of a specimen with fishbone sculpture.
19. Detail of the posterior side.

☐ 20, 21. UB 943 (sample 6749): Gum, zone I.
21. Oblique lateral view of a two-element cluster.
20. Detail of annulated outer surface. The sculpture possibly derived from resorption and subsequently coated.

☐ 23. UB 944 (sample 6811): Haggården–Marieberg, zone II.
Strongly compressed fragmentary cluster with thick coating all around.

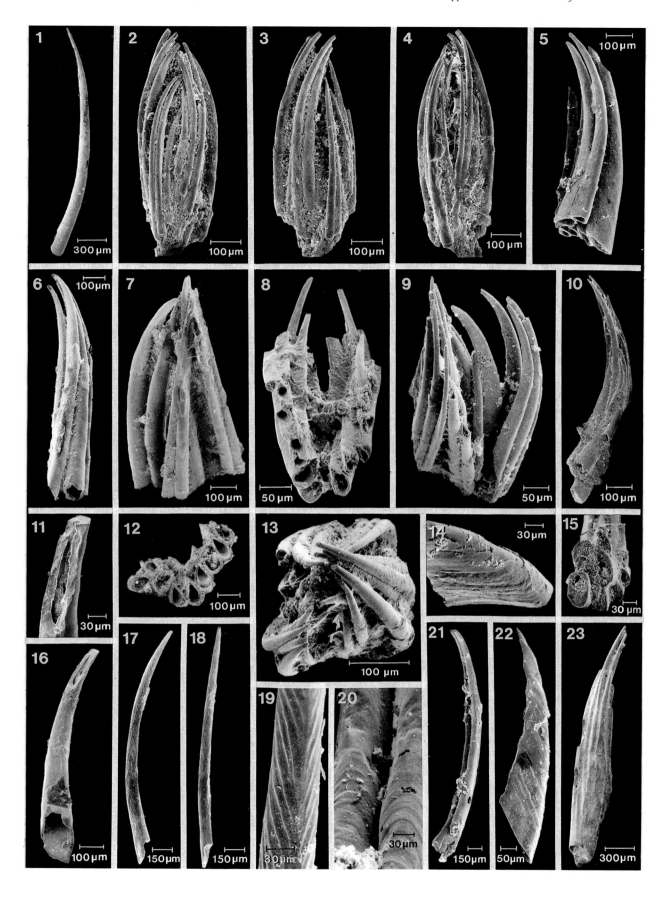

Plate 2

1–24. *Phakelodus tenuis* (Müller 1959), p. 33.

☐ 1, 8. UB 945 (sample 6409): Gum, zone I.
Lateral view and cross-section of fragmentary single specimen with prevalent microfolds.

☐ 2, 9. UB 946 (sample 6160): Stenstorp–Dala, zone Vb.
2. Lateral view, showing different layers with particularly large interspace at the tip.
9. Slightly deformed cross-section.

☐ 3, 10. UB 947 (sample 6162): Stenstorp–Dala, zone Vb.
3. Isolated sclerite exposing different layers and coating.
10. Detail of same specimen.

☐ 4, 5. UB 948 (sample 6761): Gum, zone I.
4. Lateral view of broad specimen with cancellate sculpture. Tip probably represents infilling of the basal opening.
5. Detail of sculpture.

☐ 6, 7, 11. UB 949 (sample 5951): Stenstorp–Dala, zone Vc.
6. Lateral view of tall specimen.
7. Detail of the upper portion which displays the multi-lamellar structure.
11. Cross-section.

☐ 12. UB 950 (sample 7232): S. Möckleby–Degerhamn, zone Vc.
Side view of sclerite with fine, distinct annulation. The single ribs develop from almost perpendicular to the growth axis to an increasingly oblique direction.

☐ 13, 14. UB 951 (sample 6783): Gum, zone I.
The cancellation of this specimen is formed by major ribs which are crossed by subordinate microfolds.

☐ 15, 16. UB 952 (sample 6369): Brattefors, zone V undiff.
Deformed apparatus from opposite views. Originally the sclerites have been orientated in a circle. By slight compression, parts of it appear now gently rotated against each other.

☐ 17. UB 953 (sample 5952): Stenstorp–Dala, zone Vc.
Possibly a fourth of an apparatus. Cluster corresponding to one half of the specimen illustrated in Figs. 15, 16.

☐ 18. UB 954 (sample 6761): Gum, zone I.
Fragment, exposing cancellate sculpture.

☐ 19, 20. UB 955 (sample 6762): Gum, zone I.
Lateral view of sclerite with microfolds but without major ribs perpendicular to the former.

21–24. Various clusters with different phosphatisation.

☐ 21. UB 956 (sample 990): Trolmen, zone Vc.

☐ 22. UB 957 (sample 5965): Trolmen, zone V undiff.

☐ 23. UB 958 (sample 5948): Stenstorp–Dala, zone Vb.
Possibly coprolithic.

☐ 24. UB 959 (sample 5662): Degerhamn, zone Vc.

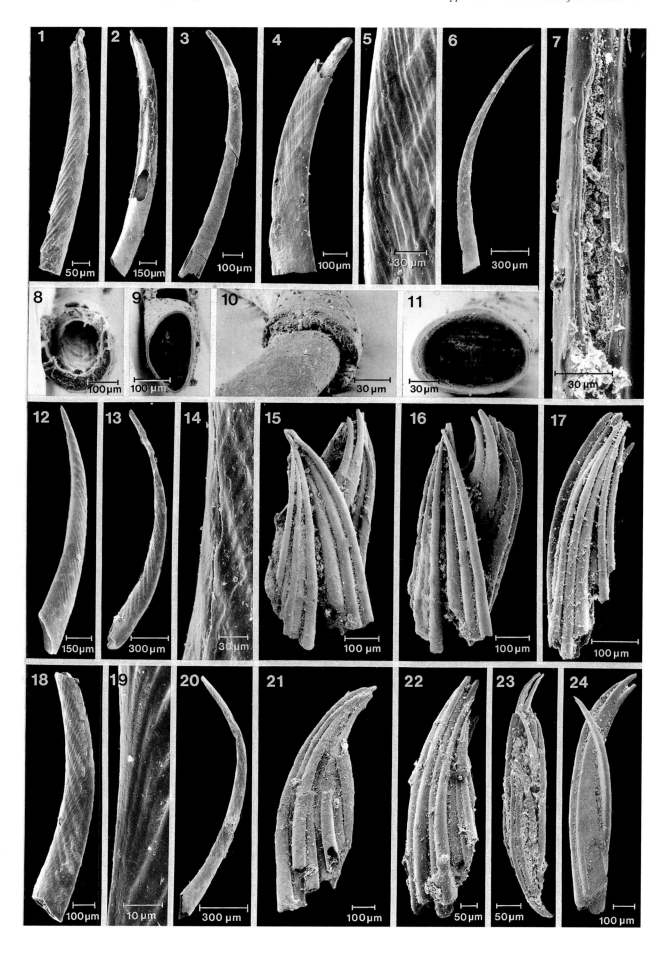

Plate 3

1–42. *Gapparodus bisulcatus* (Müller 1959), p. 25.

☐ 1, 2. UB 960 (sample 6760): Gum, zone I.
Posterior and lateral view.

☐ 3, 11. UB 961 (sample 6763): Gum, zone I.
3. Lateral view of fragment with stripe-like remains of the overlying layer.
11. Basal cross-section.

☐ 4, 5, 12. UB 962 (sample 6783): Gum, zone I.
4. Lateral view.
5. Detail of the anterior side, showing the early stages of annular resorption.
12. Cross-section above the basis.

☐ 6, 8, 14. UB 963 (sample 6768): Gum, zone I.
Lateral view with cross-section of the fractured upper end and of the basis.

☐ 7, 9, 15. UB 964 (sample 6763): Gum, zone I.
9. Side view of fragment.
7. Cross-section close to the tip, showing the extremely thick wall of the anterior side.
15. Cross-section at the basis, showing that the wall thickness decreases somewhat towards the basal rim.

☐ 10, 17. UB 965 (sample 6763): Gum, zone I.
Lateral view and fractured upper end.

☐ 13, 20, 21. UB 966 (sample 6763): Gum, zone I.
Posterior and lateral view as well as cross-section near the tip, showing initial, undifferentiated basal opening.

☐ 16, 24. UB 967 (sample 6763): Gum, zone I.
Lateral view and cross-section at the basis, exposing various layers.

☐ 18, 25. UB 968 (sample 6762): Gum, zone I.
Side view and cross-section at the basis with well-preserved anterior side.

☐ 19, 26. UB 969 (sample 6741): Klippan, zone I.

Posterior view and cross-section of basis of a thin-walled, exfoliated specimen.

☐ 22, 27. UB 970 (sample 6763): Gum, zone I.
Side view and cross-section at the basis. Both anterior and posterior side have the same wall-thickness due to loss of the outermost anterior layer.

☐ 23, 28. UB 971 (sample 6763): Gum, zone I.
Lateral view and cross-section at the basis with same phenomenon as described above.

29, 30, 37, 38. Stenstorp, Middle Cambrian.
Lateral view of fragments and cross-sections at the basis (for comparison).

☐ 29, 37. UB 972 (sample 998).

☐ 30, 38. UB 973 (sample 998).

☐ 31, 39. UB 974 (sample 6729): Backeborg, zone I.
Side view and cross-section at the basis of partly exfoliated specimen.

☐ 32. UB 975 (sample 6768): Gum, zone I.
Side view of specimen having only some stripes of the outermost anterior layer left.

☐ 33. UB 976 (sample 6753): Gum, zone I.
Lateral view, showing delicate, regular phosphatic remains on the anterior side.

☐ 34, 40. UB 977 (sample 6760): Gum, zone I.
Side view and cross-section at the basis. The latter displays different layers which are visible as different shades.

35, 36, 41, 42. Gum, zone I.
Two specimens which differ from all the others in the peculiar development of their lateral furrows.

☐ 35, 41. UB 978 (sample 6768).

☐ 36, 42. UB 979 (sample 6760).

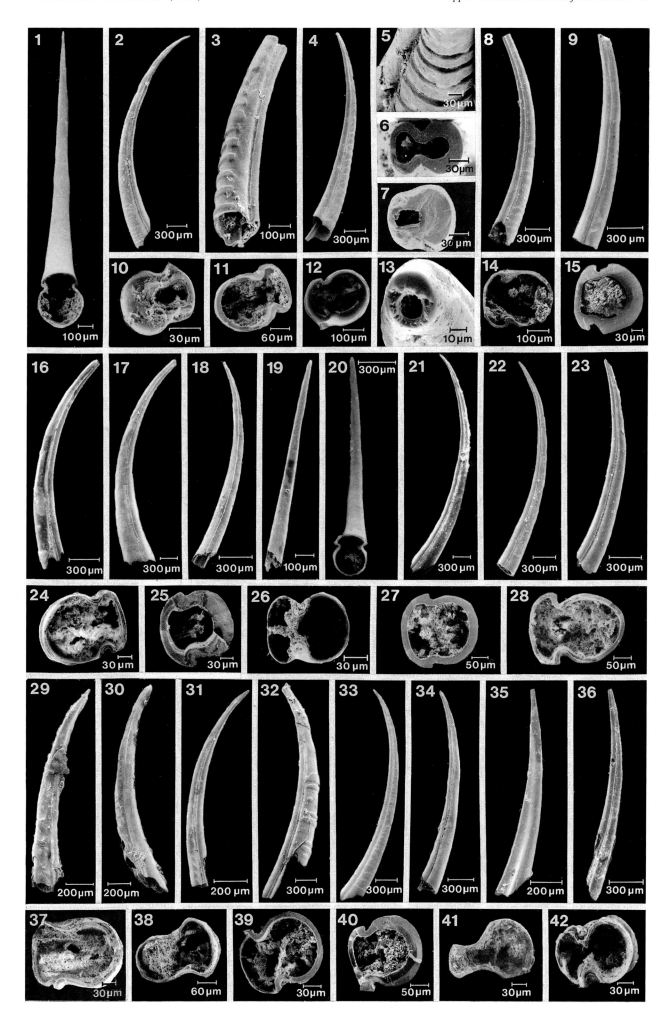

Plate 4

1–25. *Gumella cuneata* n.gen., n.sp., p. 26.

☐ 1–5. UB 980 (sample 6781): Gum, zone I. Morphotype alpha. Holotype.
1. Oblique lateral view with partly broken posterior face.
2. Posterior view.
3. Detail underneath broken posterior face, exposing a small canal.
4. Detail of the canal.
5. Cross-section at the basis.

☐ 6, 7. UB 981 (sample 955): Gudhem, zone I. Morphotype alpha.
7. Oblique lateral view with the outermost posterior face mostly broken.
6. Detail, showing the small canal.

☐ 8, 13. UB 982 (sample 6755): Gum, zone I. Morphotype alpha.
8. Posterior view. The upper part of the basal opening is exposed by fracture.
13. View of the flared basis.

☐ 9, 14, 25. UB 983 (sample 6760): Gum, zone I. Morphotype alpha.
Lateral, basal and posterior view.

☐ 10, 15. UB 984 (sample 6414): Gum, zone I. Morphotype alpha.

15. Lateral view of fragmentary specimen.
10. Detail of the upper part, showing main basal opening with joining canal.

☐ 11, 16. UB 985 (sample 6781): Gum, zone I. Morphotype alpha.
Detail of the tip and oblique posterior view.

☐ 12, 17. UB 986 (sample 6761): Gum, zone I. Morphotype alpha.
Side view of fragmentary specimen and view onto the basis.

☐ 18–20. UB 987 (sample 6768): Gum, zone I. Morphotype alpha.
Posterior view with detail of the upper portion and basal cross-section.

☐ 21, 22. UB 988 (sample 6414): Gum, zone I. Probably morphotype alpha.
21. Fragment with view onto the posterior faces of the anterior side exposing fibrous internal structure.
22. Detail.

☐ 23, 24. UB 989 (sample 6760): Gum, zone I. Morphotype alpha.
Detail of the inner surface dotted with pits, and posterior view.

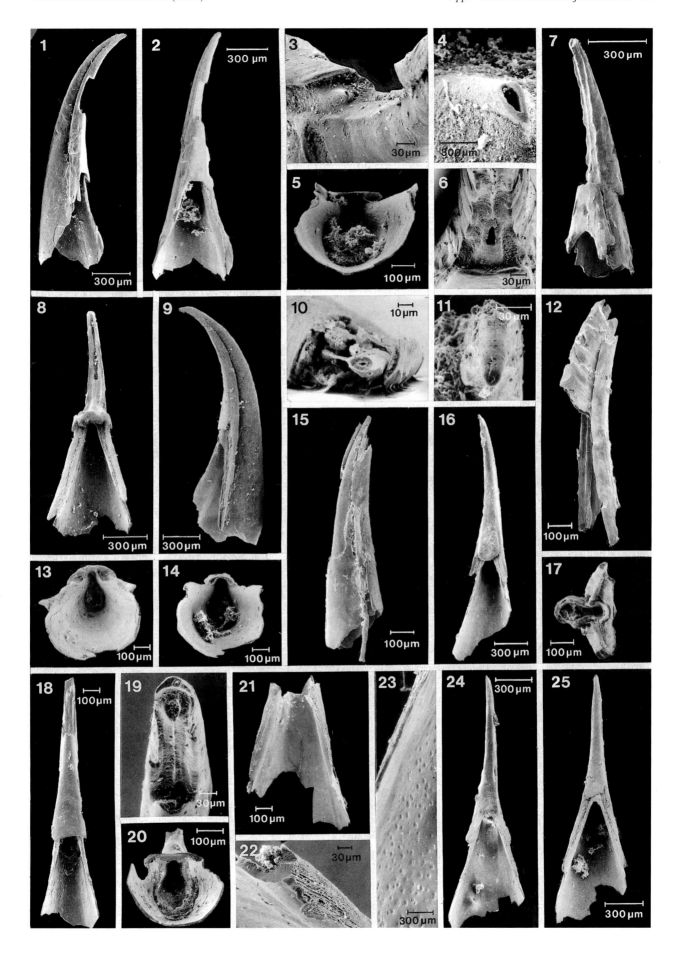

Plate 5

1–28: *Gumella cuneata* n.gen., n.sp., p. 26.

☐ 1–3. UB 990 (sample 6761): Gum, zone I. Morphotype beta.
Oblique lateral view from opposite sides and from the posterior with view into the basal opening.

☐ 4, 5. UB 991 (sample 6762): Gum, zone I. Morphotype beta.
Lateral view and detail of the exfoliated tip with exposed facets.

☐ 6, 7. UB 992 (sample 6735): Backeborg, zone I. Morphotype beta.
Basal cross-section and lateral view.

☐ 8. UB 993 (sample 6785): Gum, zone I. Morphotype beta.

☐ 9, 10. UB 994 (sample 6768): Gum, zone I. Morphotype alpha.
9. Anterior view, documenting the flaring flanks.
10. Detail of the tip. Structure possibly due to secondary phosphatisation.

☐ 11, 12. UB 995 (sample 6409): Gum, zone I. Morphotype beta.
Oblique lateral and posterior views of an extremely strongly phosphatised specimen.

☐ 13, 14. UB 996 (sample 6781): Gum, zone I. Morphotype beta.
Cross-section having several layers exposed and lateral view.

☐ 15. UB 997 (sample 6703): Stenstorp–Dala, zone Vc. Morphotype beta.
Side view of a well-preserved specimen. Anterolaterally, microfolds oblique to the growth axis are developed; posterolaterally, the fibrous structure is exposed.

☐ 16–18. UB 998 (sample 6410): Gum, zone I. Morphotype beta.
Lateral view, basal cross-section and detail of the tip.

☐ 19–21. UB 999 (sample 6409): Gum, zone I. Morphotype beta.
Lateral and posterolateral view and basal cross-section of morphotype b(?).

☐ 22–24. UB 1000 (sample 6763): Gum, zone I. Morphotype beta.
22. Oblique posterolateral view.
23. Detail of the structure close to the lateral furrow.
24. Basal cross-section.

☐ 25, 26. UB 1001 (sample 6410): Gum, zone I. Morphotype beta.
Posterolateral view and detail of the exfoliated part of the posterior side.

☐ 27. UB 1002 (sample 6734): Backeborg, zone I. Morphotype beta.
Posterior side of a widely flared specimen with view onto the broken apex.

☐ 28. UB 1003 (sample 6409): Gum, zone I. Morphotype beta.
Posterior view of much exfoliated specimen. The strongly differentiated outer layers are not preserved.

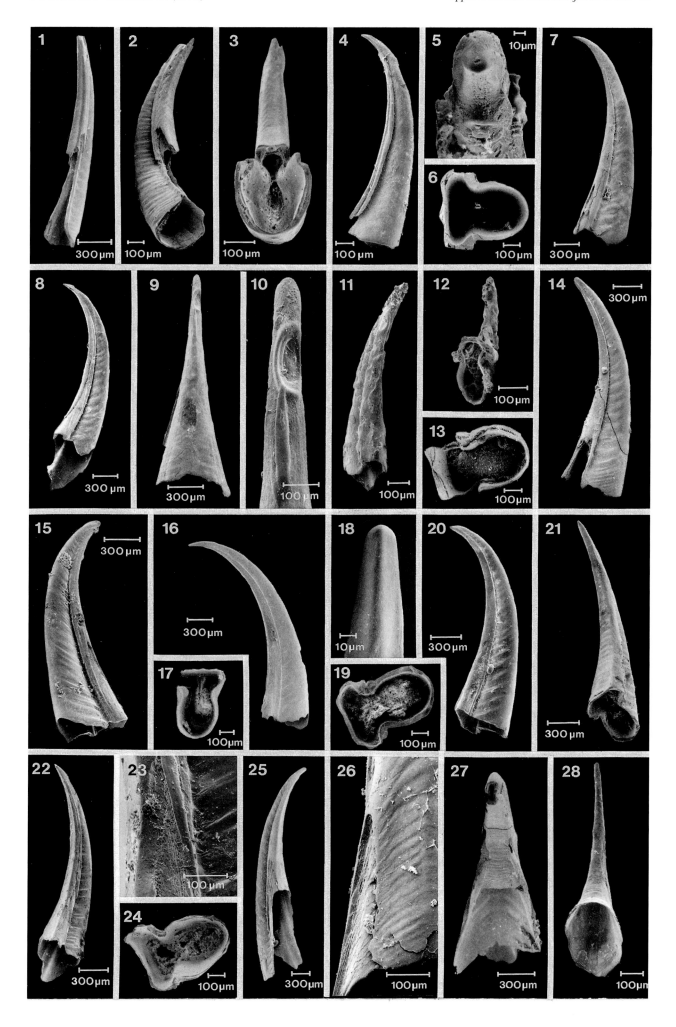

Plate 6

1–7: *Bengtsonella triangularis* n.gen., n.sp., p. 15.

☐ 1. UB 1004 (sample 6478): Degerhamn, zone Vc; ×110.
Posterior view.

☐ 2. UB 1005 (sample 6714): Stenstorp–Dala, zone Vb;
×110.
Oblique posterior view.

☐ 3–5. UB 1006 (sample 6211): Stenstorp–Dala, zone Vc;
holotype.
3. ×110. Oblique posterior view with basal opening.
4. ×140. Posterior view with basal cross-section.
5. ×140. Oblique posterior view.

☐ 6, 7. UB 1007 (sample 6414): Gum, zone Vc; ×110.
View from above and lateral view.

8–15. *Phakelodus simplex* n.sp., p. 33. all ×150.

☐ 8, 9. UB 1008 (sample 6406): Gum, zone I.
Side view and view from above.

☐ 10. UB 1009 (sample 7228): S. Möckleby–Degerhamn,
zone Vc.
Posterolateral view.

☐ 11, 12. UB 1010 (sample 6419): Trolmen, zone Vc.
Posterolateral view and view from above.

☐ 13. UB 1011 (sample 6406): Gum, zone Vc.
Posterior view.

☐ 14. UB 1012 (sample 6713): Stenstorp–Dala, zone Vc.
Holotype.
Lateral view.

☐ 15. UB 1013 (sample 5969): Trolmen, zone Vc.
View from above.

Plate 7

1–4, 6–10, 12, 13, 15, 19, 21. *Furnishina vasmerae* n.sp., p. 24, all ×80.

☐ 1. UB 1014 (sample 6760): Gum, zone I.
View from above.

☐ 2. UB 1015 (sample 6408): Gum, zone I.
Oblique anterolateral view.

☐ 3, 4. UB 1016 (sample 6768): Gum, zone I. Holotype.
View from above and oblique posterolateral view.

☐ 6, 9. UB 1017 (sample 6414): Gum, zone I.
Oblique anterior and anterolateral view.

☐ 7. UB 1018 (sample 6763): Gum, zone I.
Oblique posterior view.

☐ 8, 12. UB 1019 (sample 6760): Gum, zone I.
Posterior and lateral view.

☐ 10. UB 1020 (sample 6760): Gum, zone I.
Oblique posterior view of a slender, little differentiated specimen. Such elements have been referred to this species only on the basis of co-occurrence in the same sample.

☐ 13. UB 1021 (sample 6763): Gum, zone I.
Lateral view.

☐ 15. UB 1022 (sample 6734): Backeborg, zone I.
Posterior view.

☐ 19. UB 1023 (sample 6735): Backeborg, zone I.
Oblique posterior view.

☐ 21. UB 1024 (sample 6410): Gum, zone I.
Oblique posterior view.

5, 11, 14, 16, 17: *Furnishina alata* Szaniawski 1971, p. 16, all ×80.

☐ 5. UB 1025 (sample 6757): Gum, zone I.
Oblique posterior view.

☐ 11, 17. UB 1026 (sample 6414): Gum, zone I.
Posterior and basal view.
Note the outcropping of the growth lamellae within the basal opening.

☐ 14. UB 1027 (sample 6776): Gum, zone I.
Posterior view.

☐ 16. UB 1028 (sample 6768): Gum, zone I.
Lateral view.

18, 20: *Furnishina* cf. *alata* Szaniawski 1971, p. 16, all ×120.

☐ 18. UB 1029 (sample 994): Sätra, zone I.
Oblique posterior view.

☐ 20. UB 1030 (sample 994): Sätra, zone I.
Oblique posterior view.

Plate 8

1–6, 9: *Furnishina bicarinata* Müller 1959, p. 17, all ×50.

☐ 1, 2. UB 1031 (sample 7217): S. Möckleby–Degerhamn, zone Vc.
Lateral and posterior view.

☐ 3. UB 1032 (sample 7217): S. Möckleby–Degerhamn, zone Vc.
Oblique posterior view.

☐ 4. UB 1033 (sample 6776): Gum, zone I.
Oblique posterior view.

☐ 5. UB 1034 (sample 6796): Sätra, zone I.
Posterior view.

☐ 6. UB 1035 (sample 6227): Ödegården, zone Va.
Posterior view.

☐ 9. UB 1036 (sample 6737): Backeborg, zone I.
Oblique side view. The outer surface exhibits a fine annulation.

7, 8, 10–22: *Furnishina ovata* n.sp., p. 22, all ×120.

☐ 7, 10. UB 1037 (sample 6162): Stenstorp–Dala, zone Vc.
Lateral and posterior view.

☐ 8. UB 1038 (sample 6238): Ödegården, zone Vb.
Posterior view.

☐ 11, 15. UB 1039 (sample 5650): Grönhögen, zone Vc.
Anterior view and view from above. Specimen with thin coating, partly exfoliated.

☐ 14. UB 1040 (sample 6323): Tomten, zone Vc.
Oblique view from above.

☐ 12, 13. UB 1041 (sample 6186): Stenstorp–Dala, zone Vb.
Oblique view from above and posterolateral view.

☐ 16, 19. UB 1042 (sample 5951): Stenstorp–Dala, zone Vc.
Holotype.
Posterior and lateral view.

☐ 17. UB 1043 (sample 6216): Stenstorp–Dala, zone Vc.
Posterior view.

☐ 18. UB 1044 (sample 6164): Stenstorp–Dala, zone Vc.
Posterior view.

☐ 20. UB 1045 (sample 5672): S. Möckleby, zone Vb or c.
Posterior view.

☐ 21. UB 1046 (sample 6341): Skår, zone Vc.
Posterior view.

☐ 22. UB 1047 (sample 7232): S. Möckleby–Degerhamn, zone Vc.
Posterior view.

Plate 9

1–13: *Furnishina quadrata* Müller 1959, p. 23.

☐ 1. UB 1048 (sample 6172): Stenstorp–Dala, zone Vb; ×100.
Posterolateral view.

☐ 2, 3. UB 1049 (sample 6793): Stubbegården, zone Vc; ×100.
Lateral view and view from above.

☐ 4. UB 1050 (sample 6802): Haggården–Marieberg, zone II; ×80.
Posterolateral view.

☐ 5. UB 1051 (sample 6806): Haggården–Marieberg, zone II; ×80.
Posterior view.

☐ 6. UB 1052 (sample 6172): Stenstorp–Dala, zone Vb or c; ×80.

Posterior view.

☐ 7, 9. UB 1053 (sample 6792): Stubbegården, zone II; ×65.
Posterior view and oblique view from above.

☐ 8. UB 1054 (sample 6810): Toreborg, zone II; ×65.
View from above.

☐ 10, 13. UB 1055 (sample 6172): Stenstorp–Dala, zone Vb; ×65.
View from above and lateral view.

☐ 11. UB 1056 (sample 6802): Haggården–Marieberg, zone II; ×80.
Oblique posterolateral view.

☐ 12. UB 1057 (sample 6402): Haggården–Marieberg, zone II; ×80.
Posterior view.

Plate 10

1–19: *Furnishina asymmetrica* Müller 1959, p. 17, all ×60.

☐ 1, 2, 5. UB 1058 (sample 6763): Gum, zone I.
Lateral, posterior, and basal view.

☐ 3, 7. UB 1059 (sample 6768): Gum, zone I.
Posterior and basal view. Inner side with beautifully exposed growth lamellae and pitted surface.

☐ 4. UB 1060 (sample 6763): Gum, zone I.
Posterior view. Note the scalloped, well-preserved lower rim.

☐ 6. UB 1061 (sample 6777): Gum, zone I.
Posterior view.

☐ 8. UB 1062 (sample 6776): Gum, zone I.
Posterior view.

☐ 9. UB 1063 (sample 6414): Gum, zone I.
Posterior view.

☐ 10. UB 1064 (sample 6412): Gum, zone I.
Oblique posterior view.

☐ 11. UB 1065 (sample 6735): Backeborg, zone I.
Posterior view.

☐ 12, 13. UB 1066 (sample 6417): Gum, zone I.
Lateral and oblique posterior view. Anterolateral laminae with regularly arranged small ribs (see also Figs. 10, 19).

☐ 14. UB 1067 (sample 6763): Gum, zone I.
Posterior view.

☐ 15. UB 1068 (sample 6414): Gum, zone I.
View from above. Anterior side with well-developed carina (see also Fig. 17).

☐ 16. UB 1069 (sample 6735): Backeborg, zone I.
Posterior view. The outcropping growth lamellae within the basal opening are exaggerated by phosphatic coating.

☐ 17. UB 1070 (sample 6163): Stenstorp–Dala, zone Vb.
Oblique lateral view.

☐ 18. UB 1071 (sample 6755): Gum, zone I.
Oblique posterior view.

☐ 19. UB 1072 (sample 6760): Gum, zone I.
Posterior view.

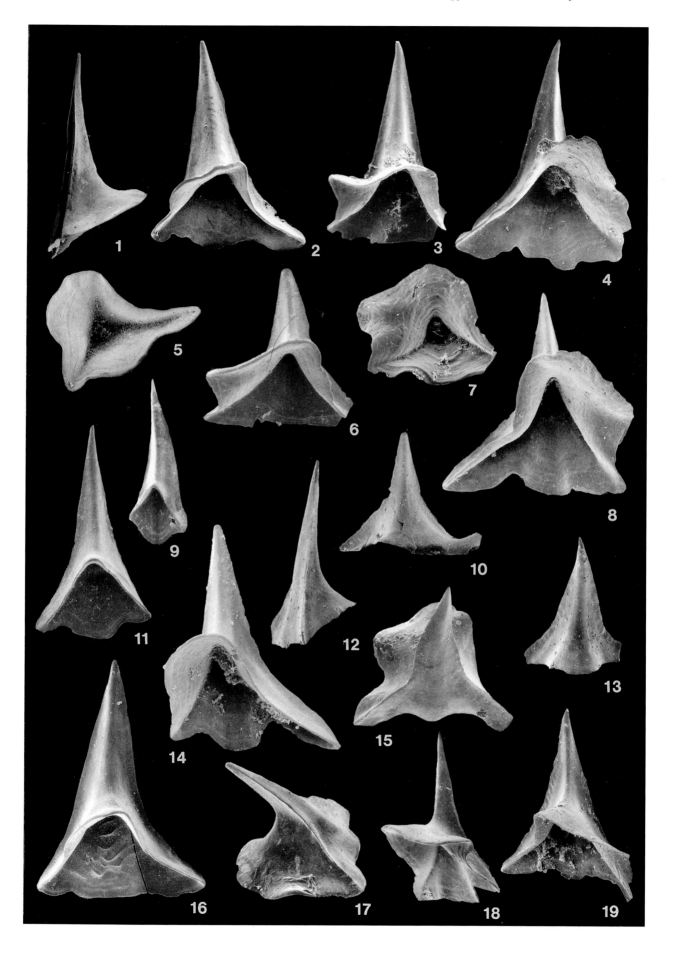

Plate 11

1–9, 11: *Furnishina polonica* Szaniawski 1971, p. 22, all ×85.

☐ 1. UB 1073 (sample 6734): Backeborg, zone I.
Oblique view from above.

☐ 2. UB 1074 (sample 6754): Gum, zone I.
Posterior view.

☐ 3. UB 1075 (sample 6417): Gum, zone I.
Oblique view from above.

☐ 4. UB 1076 (sample 6761): Gum, zone I.
Posterior view.

☐ 5. UB 1077 (sample 6763): Gum, zone I.
Posterior view.

☐ 6, 8: UB 1078 (sample 6762): Gum, zone I.
Posterior and basal view.

☐ 7. UB 1079 (sample 6755): Gum, zone I.
Oblique posterior view.

☐ 9, 11. UB 1080 (sample 6772): Gum, zone I.
Lateral view and view from above.

10, 12. *Furnishina longibasis* Bednarczyk 1979, p. 21.

☐ 10, 12. UB 1081 (sample 6404): Haggården–Marieberg,
zone II; ×85.
Oblique lateral view and view from above.

Plate 12

1, 2, 6, 8, 12–14, 18. *Furnishina kranzae* n.sp., p. 21, all ×60.

☐ 1. UB 1082 (sample 6794): Stubbegården, zone II.
Posterior view.

☐ 2, 6. UB 1083 (sample 6806): Österplana, zone II.
Posterior and basal view.

☐ 8, 13. UB 1084 (sample 6806): Österplana, zone II. Holotype.
Posterior and basal view.

☐ 12. UB 1085 (sample 6806): Haggården–Marieberg, zone II.
Anterolateral view.

☐ 14, 18. UB 1086 (sample 6806): Haggården–Marieberg, zone II.
Posterior and basal view.

3, 4, 7, 9, 10, 15–17, 19. *Furnishina primitiva* Müller 1959, p. 23, all ×60.

☐ 3, 7. UB 1087 (sample 5650): Grönhögen, zone Vc.
Posterior and basal view.

☐ 4. UB 1088 (sample 6831): Trolmen, zone V undiff.
Oblique lateral view.

☐ 5, 10, 11. UB 1089 sample 5650): Grönhögen, zone Vc.
Posterior, lateral, and basal view.

☐ 9, 15. UB 1090 (sample 5659): Degerhamn, zone Vc.
Posterior and basal view.

☐ 16. UB 1091 (sample 5650): Grönhögen, zone Vc.
Oblique anterior view.

☐ 17. UB 1092 (sample 5650): Grönhögen, zone Vc.
Oblique posterior view.

☐ 19. UB 1093 (sample 5659): Degerhamn, zone Vc.
Lateral view.

20, 21. *Acodus cambricus* Nogami 1966, p. 51, all ×160.

☐ 20, 21. UB 1094 (sample 6414): Gum, zone I.
View from above and lateral view. Euconodont with some similarity to *Furnishina* in its gross morphology.

Plate 13

1–7, 11, 12. *Furnishina furnishi* Müller 1959, p. 17, all ×80.

☐ 1, 2. UB 1095 (sample 6160): Stenstorp–Dala, zone Vb. Posterior and lateral view.

☐ 3, 4. UB 1096 (sample 6209): Stenstorp–Dala, zone Vc. Posterior and lateral view.

☐ 5. UB 1097 (sample 5650): Grönhögen, zone Vc. Oblique posterior view. Semispherical structures of unknown origin are attached to both posterolateral faces.

☐ 6. UB 1098 (sample 6172): Stenstorp–Dala, zone Vb. Posterior view of an early growth stage.

☐ 7. UB 1099 (sample 6160): Stenstorp–Dala, zone Vb. Oblique posterior view.

☐ 11. UB 1100 (sample 6369): Brattefors, zone V undiff. Posterior view of cluster of three small sclerites.

☐ 12. UB 1101 (sample 6227): Ödegården, zone Va. Posterior view.

8–10, 13, 14, 16, 17, 21. *Furnishina gossmannae* n.sp., p. 20, all ×80.

☐ 8, 9. UB 1102 (sample 6166): Stenstorp–Dala, zone Vb. Oblique lateral and posterior view.

☐ 10, 16, 17. UB 1103 (sample 6238): Ödegården, zone Vc. Holotype. Posterior, basal, and lateral view.

☐ 13. UB 1104 (sample 6228): Ödegården, zone Vb?. Posterior view of an early growth stage.

☐ 14. UB 1105 (sample 6208): Stenstorp–Dala, zone Vc. Oblique posterior view of an early growth stage.

☐ 21. UB 1106 (sample 1000): Grönhögen, zone III. Posterior view.

15, 18, 20, 22–25. *Furnishina curvata* n.sp., p. 17, all ×115.

☐ 15, 18. UB 1107 (sample 6172): Stenstorp–Dala, zone Vb. Oblique view from above and lateral view.

☐ 20. UB 1108 (sample 6228): Ödegården, zone Vb?. Anterolateral view.

☐ 22. UB 1109 (sample 6186): Stenstorp–Dala, zone Vb. Lateral view.

☐ 23, 24. UB 1110 (sample 6228): Ödegården, zone Vb?. Posterior view and basal cross-section of an early growth stage.

☐ 25. UB 1111 (sample 6228): Ödegården, zone Vb?. Holotype. Lateral view.

19, 26. *Furnishina* sp. indet., p. 25.

☐ 19, 26. UB 1112 (sample 6415): Gum, zone I; ×150. 19. Lateral view, showing bumpy posterior keel. 26. Basal view, showing coarse growth lamellae; possibly bundles of the latter.

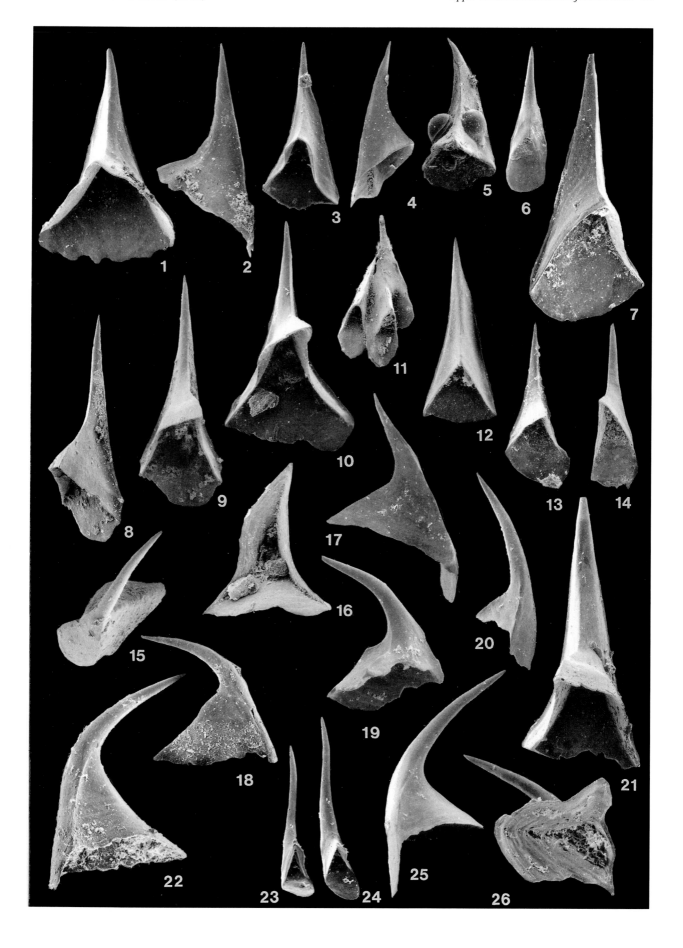

Plate 14

1. *Furnishina?* sp., p. 25.

☐ 1. UB 1113 (sample 6747): St. Stolan, zone I; ×150.
Posterior view of a single specimen with overall similarity
to *Furnishina tortilis* except for a secondary lateral denticle.

2–19. *Furnishina tortilis* (Müller 1959), p. 24.

☐ 2. UB 1114 (sample 6735): Ekebacka, zone II; ×40.
Posterior view.

☐ 3. UB 1115 (sample 6414): Gum, zone I; ×60.
Posterior view.

☐ 4, 5. UB 1116 (sample 6365): St. Stolan, zone I; ×40.
Anterolateral and oblique posterior view.

☐ 6, 12. UB 1117 (sample 6404): Haggården–Marieberg,
zone II; ×40.
Posterior and lateral view.

☐ 7. UB 1118 (sample 6768): Gum, zone I; ×40.
Posterolateral view.

☐ 8, 13. UB 1119 (sample 6750): Gum, zone I.

8. Posterior view; ×40.
13. Basal cross-section; ×60.

☐ 9. UB 1120 (sample 6777): Gum, zone I; ×40.
Posterior view.

☐ 10, 11. UB 1121 (sample 6404): Haggården–Marieberg,
zone II; ×40.
Posterior and lateral view.

☐ 15. UB 1122 (sample 6776): Gum, zone I; ×40.
Oblique anterolateral view. Deformed specimen with
broken tip exposing the growth lamellae.

14, 16–19. Single specimens with extremely large, highly
differentiated posterior side.

☐ 14. UB 1123 (sample 6741): Klippan, zone I; ×40.
Oblique lateral view.

☐ 16, 18. UB 1124 (sample 6760): Gum, zone I; ×40.
Oblique lateral view and view from above.

☐ 17, 19. UB 1125 (sample 6763): Gum, zone I; ×40.
Basal and oblique lateral view.

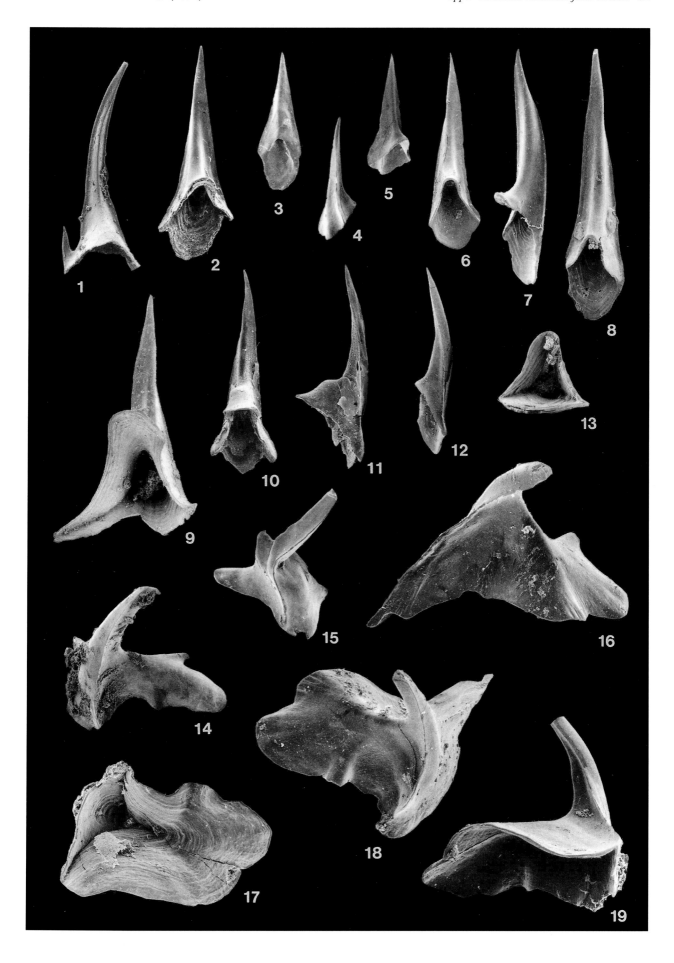

Plate 15

1–5, 7–10, 16. *Furnishina kleithria* n.sp, p. 21, all ×130.

☐ 1. UB 1126 (sample 6763): Gum, zone I.
Posterior view of specimen with well-developed antero-lateral laminae.

☐ 2, 7. UB 1127 (sample 6414): Gum, zone I.
Posterior and basal view.

☐ 3, 8. UB 1128 (sample 6409): Gum, zone I. Holotype.
Lateral and posterior view.

☐ 4, 9. UB 1129 (sample 6414): Gum, zone I.
Posterolateral and oblique anterior view.

☐ 5, 10. UB 1130 (sample 6739): Kakeled, zone I.
Posterior and basal view.

☐ 16. UB 1131 (sample 6748): Gum, zone I.
Posterior view.

6, 11–15, 17–20. *Furnishina sinuata* n.sp., p. 24.

☐ 6, 11. UB 1132 (sample 6232): Ödegården, zone V; ×100.
Posterior and basal view.

☐ 12. UB 1133 (sample 6369): Brattefors, zone V undiff.; ×50.
Lateral view of broken specimen. Note pitted surface of the inner side.

☐ 13, 14. UB 1134 (sample 7232): S. Möckleby–Deger-hamn, zone Vc; ×70.
Oblique posterior and anterolateral view.

☐ 15, 20. UB 1135 (sample 6482): Mörbylilla–Albrunna, zone V undiff.; ×70.
Posterior and anterolateral view.

☐ 17–19. UB 1136 (sample 6213): Stenstorp–Dala, zone Vb; ×70. Holotype.
Oblique posterior, lateral and basal view.

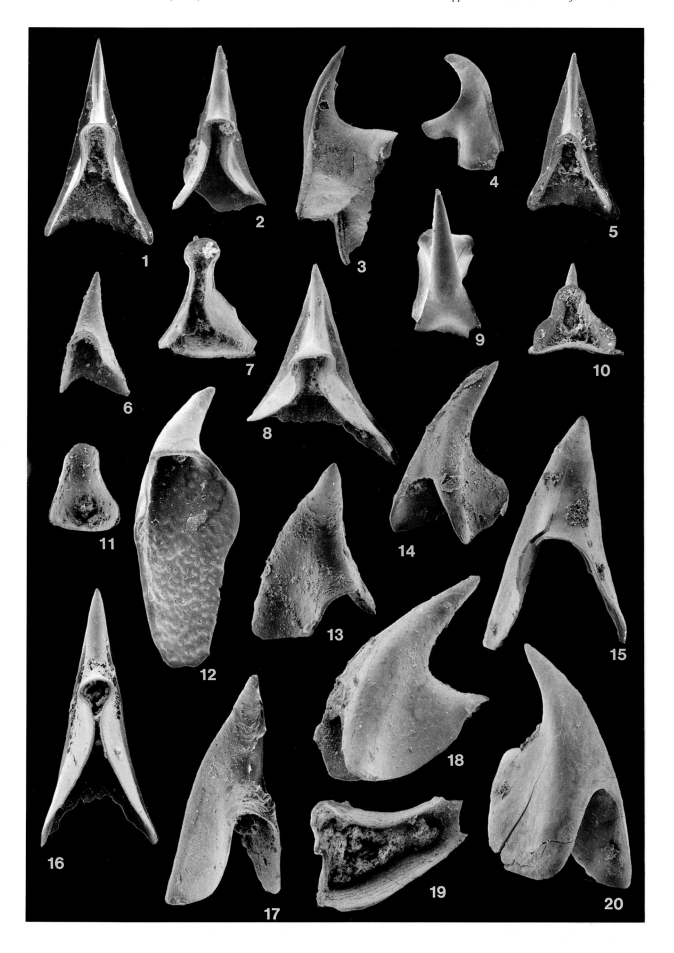

Plate 16

1–13, 15. *Furnishina gladiata* n.sp, p. 20.

☐ 1. UB 1137 (sample 6772): Gum, zone I; ×100.
   Posterior view.

☐ 2–4. UB 1138 (sample 6783): Gum, zone I; ×90.
   Right lateral, posterior, and left lateral view.

☐ 5, 8. UB 1139 (sample 6752): Gum, zone I; ×90.
   Oblique posterior and basal view.

☐ 6, 9. UB 1140 (sample 6760): Gum, zone I; ×100.
   Oblique posterolateral and anterior view.

☐ 7. UB 1141 (sample 994): Haggården–Marieberg, zone
   I; ×100.
   Basal view.

☐ 10, 13. UB 1142 (sample 6409): Gum, zone I; ×100.
   Oblique anterior view from above and lateral view.

☐ 11, 12, 15. UB 1143 (sample 6417): Gum, zone I; ×100.
   Holotype.
   Posterior, lateral, and basal view..

14. *Furnishina* sp. aff. *gladiata*, p. 20.

☐ 14. UB 1144 (sample 6746): St. Stolan, zone I; ×100.
   Posterior view.

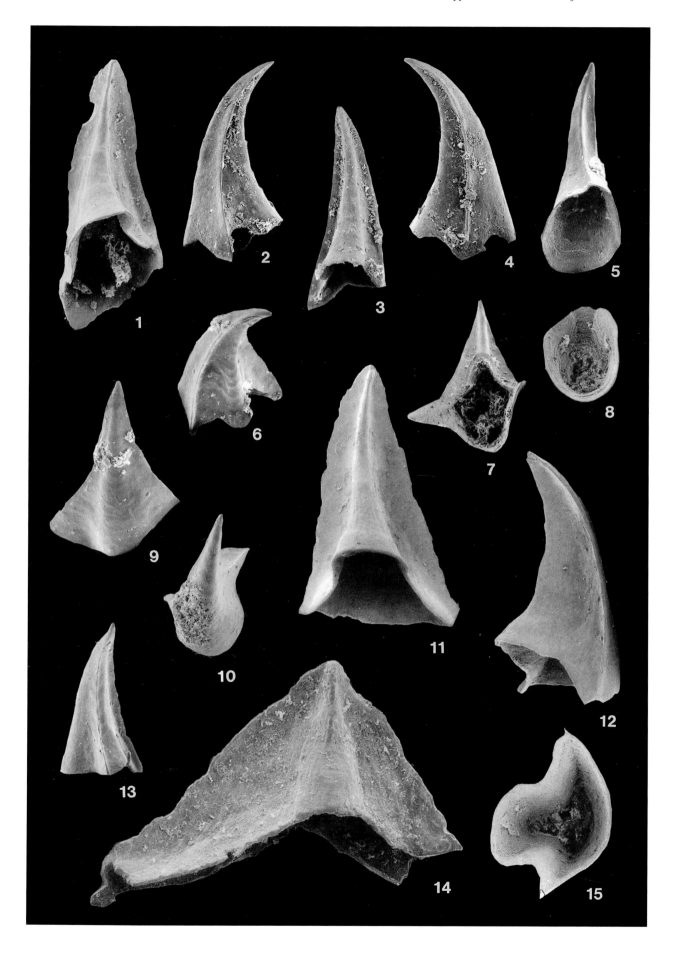

Plate 17

1–8, 10, 12, 13. *Furnishina rara* (Müller 1959), p. 23.

☐ 1, 2. UB 1145 (sample 992): Grönhögen, zone III; ×90.
Oblique posterolateral view and oblique anterolateral
view from above.

☐ 3. UB 1146 (sample 6763): Gum, zone I; ×90.
Lateral view.

☐ 4. UB 1147 (sample 6763): Gum, zone I; ×90.
Oblique posterior view.

☐ 5. UB 1148 (sample 6762): Backeborg, zone I; ×90.
Lateral view.

☐ 6. UB 1149 (sample 6735): Gum, zone I; ×65.
Oblique lateral view with pitted inner surface.

☐ 7. UB 1150 (sample 6763): Gum, zone I; ×90.
Oblique posterior view.

☐ 8. UB 1151 (sample 6763): Gum, zone I; ×90.
Oblique posterior view.

☐ 10. UB 1152 (sample 6404): Haggården–Marieberg,
zone II; ×90.
Posterior view.

☐ 12. UB 1153 (sample 6802): Toreborg, zone II; ×120.
Oblique posterior view of specimen broken longitudinally in the median plane. The small triangular body

possibly represents the shrunken lining of the basal
opening.

☐ 13. UB 1154 (sample 6811): Gum, zone I; ×90.
Oblique posterior view.

9, 11, 14–20. *Furnishina mira* n.sp., p. 22, all ×130.

☐ 9. UB 1155 (sample 6408): Haggården–Marieberg, zone
II.
View from above.

☐ 11. UB 1156 (sample 6414): Gum, zone I.
Oblique posterior view.

☐ 14, 15. UB 1157 (sample 6417): Gum, zone I.
Oblique view from above and anterolateral view.

☐ 16. UB 1158 (sample 6748): Gum, zone I.
Anterolateral view from above.

☐ 17. UB 1159 (sample 6777): Gum, zone I.
Oblique posterolateral view.

☐ 18, 19. UB 1160 (sample 6771): Gum, zone I.
Oblique posterolateral and lateral view.

☐ 20. UB 1161 (sample 6409): Gum, zone I. Holotype.
Oblique view from above.

Plate 18

1–4, 6–13, 15, 17. *Muellerodus cambricus* (Müller 1959), p. 29, all ×90.

☐ 1, 8. UB 1162 (sample 6802): Haggården–Marieberg, zone II.
Lateral and posterior view.

☐ 2. UB 1163 (sample 6755): Gum, zone I.
Oblique posterior view.

☐ 3. UB 1164 (sample 6238): Ödegården, zone Vb.
Posterior view.

☐ 4. UB 1165 (sample 6783): Gum, zone I.
Lateral view.

☐ 6, 11. UB 1166 (sample 6404): Haggården–Marieberg, zone II.
Lateral and basal view.

☐ 7. UB 1167 (sample 6783): Gum, zone I.
Posterior view.

☐ 9, 10. UB 1168 (sample 6753): Gum, zone I.
Oblique lateral and basal view of a large sclerite. The bigger flank is typically excavated.

☐ 12. UB 1169 (sample 6783): Gum, zone I.
Posterior view.

☐ 13. UB 1170 (sample 6768): Gum, zone I.
Posterolateral view.

☐ 15. UB 1171 (sample 6776): Gum, zone I.
Posterior view.

☐ 17. UB 1172 (sample 6784): Gum, zone I; ×11.
Lateral view of fragment, showing solid lateral costa. The wall thickness decreases considerably towards the basis.

5, 14, 16, 18–21. *Muellerodus pomeranensis.* Szaniawski 1971, p. 30.

☐ 5. UB 1173 (sample 6772): Gum, zone I. ×110.
Oblique posterior view.

☐ 14, 18. UB 1174 (sample 6417): Gum, zone I; ×90.
View from above and side view of specimen with unusually developed, trilobate posterior side.

☐ 16, 19. UB 1175 (sample 6760): Gum, zone I; ×110.
Posterior and basal view.

☐ 20. UB 1176 (sample 6785): Gum, zone I; ×110.
Posterior view. Note the vertical ribs on the lamina (see also Fig. 14).

☐ 21. UB 1177 (sample 6810): Toreborg, zone II; ×90.
Posterior view.

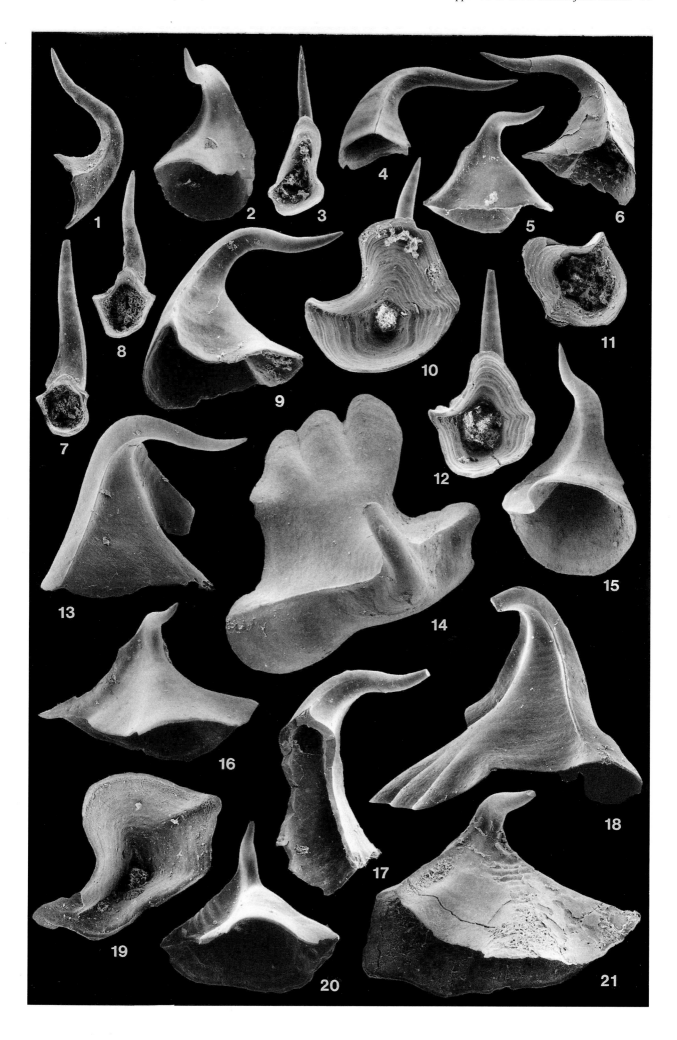

Plate 19

1–9, 11. *Muellerodus guttulus* n.sp., p. 29, all ×180.

☐ 1, 4. UB 1178 (sample 6784): Gum, zone I.
Basal and posterior view.

☐ 2, 3. UB 1179 (sample 6731): Backeborg, zone I.
Posterior and posterolateral view.

☐ 5, 8. UB 1180 (sample 6752): Gum, zone I. Holotype.
Posterior and basal view. Note the distinct laminae on
both sides.

☐ 6, 9. UB 1181 (sample 6761): Gum, zone I.
Posterior and basal view.

☐ 7. UB 1182 (sample 6414): Gum, zone I.
Posterior view.

☐ 11. UB 1183 (sample 6796): Gum, zone I.
Posterior view.

10, 12–19. *Muellerodus subsymmetricus* n.sp., p. 30.

☐ 10, 12. UB 1184 (sample 6748): Gum, zone I; ×180.
Oblique posterolateral and posterior view.

☐ 13, 14. UB 1185 (sample 6404): Haggården–Marieberg,
zone II; ×180.
Cluster of four sclerites, not in the original orientation
to each other.

☐ 15, 19. UB 1186 (sample 6783): Gum, zone I; ×110.
Holotype.
Lateral and basal view.

☐ 16. UB 1187 (sample 6413): Gum, zone I; ×180.
Posterolateral view.

☐ 17. UB 1188 (sample 6414): Gum, zone I; ×180.
Posterior view.

☐ 18. UB 1189 (sample 6414): Gum, zone I; ×180.
View from above.

Plate 20

1–13. *Muellerodus? oelandicus* (Müller 1959), p. 29, all ×150.

☐ 1. UB 1190 (sample 6228): Ödegården, zone Vb?.
Lateral view.

☐ 2. UB 1191 (sample 6812): Gössäter, zone I.
Oblique lateral view.

☐ 3, 5. UB 1192 (sample 6804): Haggården–Marieberg,
zone II.
Posterolateral and basal view.

☐ 4. UB 1193 (sample 7232): S. Möckleby–Degerhamn,
zone Vc.
Posterolateral view.

☐ 6. UB 1194 (sample 5949): Stenstorp–Dala, zone Vc.

Lateral view.

☐ 7. UB 1195 (sample 6410): Gum, zone I.
Anterior view.

☐ 8. UB 1196 (sample 6812): Gössäter, zone I.
Posterior view.

☐ 9. UB 1197 (sample 6267): Smedsgården–Stutagården,
zone Vc.
Lateral view.

☐ 10, 11. UB 1198 (sample 6238): Ödegården, zone Vb.
Oblique anterior view from above and lateral view.

☐ 12, 13. UB 1199 (sample 6219): Stenstorp–Dala, zone Vc.
Basal and lateral view.

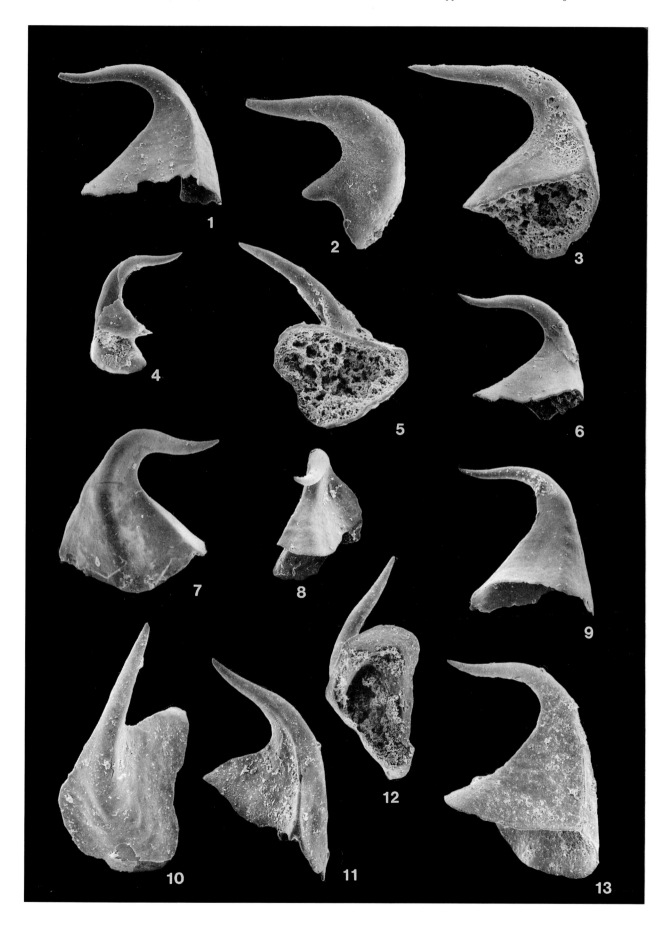

Plate 21

1–6. *Nogamiconus sinenesis* (Nogami 1966), p. 31, all ×100.

☐ 1, 2. UB 1200 (sample 6739): Klippan, zone I.
Posterolateral and posterior view.

☐ 3. UB 1201 (sample 6771): Gum, zone I.
Lateral view.

☐ 4, 5. UB 1202 (sample no 6771): Gum, zone I.
Lateral and basal view.

☐ 6. UB 1203 (sample 6771): Gum, zone I.
Oblique posterior view.

7, 8, 10. *Nogamiconus* sp., p. 32, all ×70.

☐ 7. UB 1204 (sample 5672): S. Möckleby, zone Vb.
Posterolateral view.

☐ 8. UB 1205 (sample 6238): Ödegården, zone Vc.
Lateral view.

☐ 10. UB 1206 (sample 5672): Stenstorp–Dala, zone Vb.
Lateral view.

9, 11–23. *Nogamiconus falcifer* n.sp., p. 30, all ×70.

☐ 9. UB 1207 (sample 6141): S. Möckleby, zone Vb.
Morphotype alpha.
Lateral view.

☐ 11. UB 1208 (sample 6295): Ranstadsverket, zone Vb.
Morphotype alpha.
Lateral view. Semispherical structures attached to the
outer surface, as in *Furnishina furnishi* (Pl. 13:5).

☐ 12, 15. UB 1209 (sample 990): Trolmen, zone Vc.
Morphotype beta. Holotype.
Lateral and basal view.

☐ 13. UB 1210 (sample 6186): Stenstorp–Dala, zone Vb.
Morphotype alpha.
Oblique lateral view.

☐ 14. UB 1211 (sample 7231): S. Möckleby–Degerhamn,
zone Vc. Morphotype alpha.
Lateral view.

☐ 16. UB 1212 (sample 5740): Fehmarn, zone V undiff.
Morphotype alpha.
Cluster of three specimens which are attached to each
other in opposite directions.

☐ 17. UB 1213 (sample 6718): Stenstorp–Dala, zone Vb.
Morphotype beta.
Lateral view.

☐ 18. UB 1214 (sample 5667): Degerhamn, zone Vc.
Morphotype beta.
Lateral view.

☐ 19. UB 1215 (sample 5951): Stenstorp–Dala, zone Vc.
Morphotype alpha.
Oblique basal view.

☐ 20, 22. UB 1216 (sample 6234): Ödegården, zone Vb.
Morphotype beta.
View from above and lateral view.

☐ 21. UB 1217 (sample 6142): Stenstorp–Dala, zone Vb.
Morphotype beta.
Lateral view.

☐ 23. UB 1218 (sample 5946): Stenstorp–Dala, zone Vb.
Morphotype beta.
Lateral view.

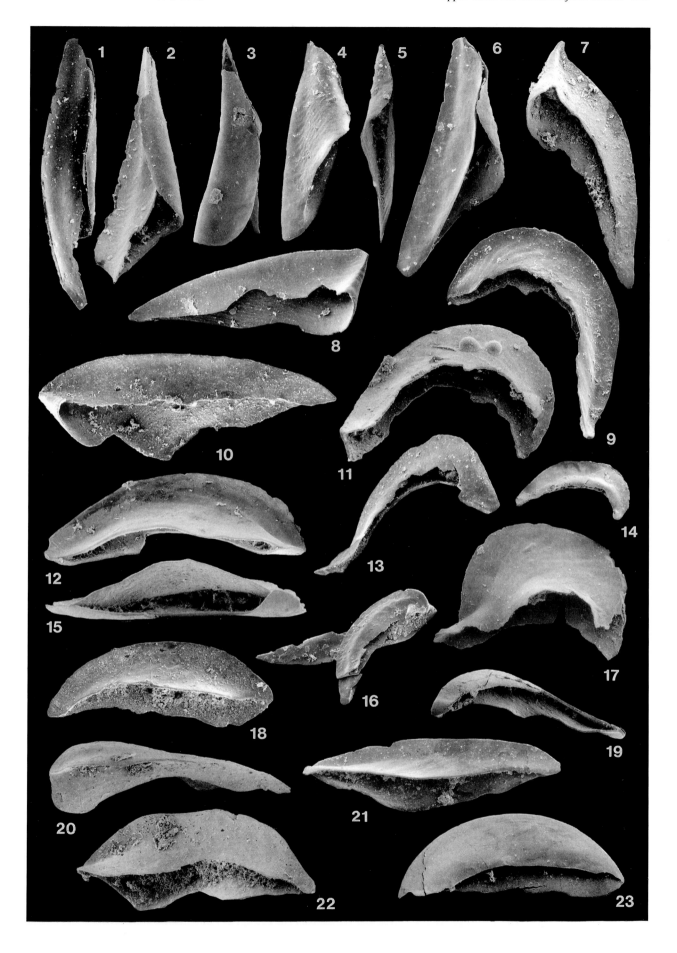

Plate 22

1, 2, 6. *Proacodus* sp., p. 35, all ×100.

☐ 1, 6. UB 1219 (sample 5652): Grönhögen, zone Vc.
Lateral and posterior view.

☐ 2. UB 1220 (sample 7231): S. Möckleby–Degerhamn,
zone Vc.
Oblique posterior view.

3–5, 7–11. *Proacodus pulcherus* (An 1982), p. 34, all ×100.

☐ 3, 4. UB 1221 (sample 7232): S. Möckleby–Degerhamn,
zone Vc.
Basal and posterior view.

☐ 5. UB 1222 (sample 6422): Trolmen, zone Vc.
Oblique posterior view.

☐ 7, 10. UB 1223 (sample 5657): Degerhamn, zone Vc.
Posterior and basal view.

☐ 8. UB 1224 (sample 5948): Stenstorp–Dala, zone Vb.
Oblique basal view.

☐ 9. UB 1225 (sample 6421): Trolmen, zone Vc.
Posterior view.

☐ 11. UB 1226 (sample 5967): Trolmen, zone V undiff.
Oblique posterior view.

12–23. *Proacodus obliquus* Müller 1959, p. 34.

☐ 12. UB 1227 (sample 6406): Gum, zone Vc; ×100.

Oblique posterior view.

☐ 13. UB 1228 (sample 5951): Stenstorp–Dala, zone Vc;
×100.
Lateral view.

☐ 14, 15. UB 1229 (sample 5951): Stenstorp–Dala, zone Vc;
×115.
Posterior and basal view.

☐ 16, 18. UB 1230 (sample 5972): S. Möckleby, zone Vb or
c; ×100.
Basal and posterior view.

☐ 17. UB 1231 (sample 6406): Gum, zone Vc; ×100.
Anterior view from above.

☐ 19. UB 1232 (sample 6341): Skår, zone Vc; ×100.
Posterior view.

☐ 20, 21. UB 1233 (sample 5949): Stenstorp–Dala, zone Vb;
×100.
Anterior view and posterior view from above.

☐ 22. UB 1234 (sample 6406): Gum, zone Vc; ×100.
Posterior view.

☐ 23. UB 1235 (sample 5944): Stenstorp–Dala, zone Vb;
×100.
Anterior view.

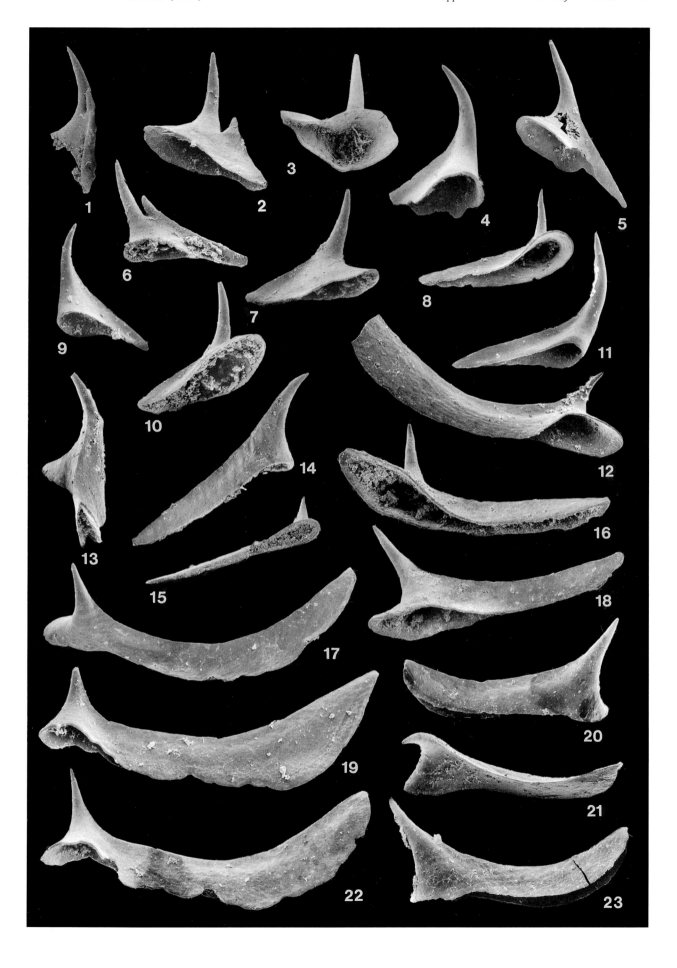

Plate 23

1–10, 14, 15, 18–20, 22. *Problematoconites perforatus* Müller 1959 emend., p. 36.

☐ 1. UB 1236 (sample 918): Grönhögen, zone Vc; ×35.
Lateral view.

☐ 2. UB 1237 (sample 7219): S. Möckleby–Degerhamn, zone Vc; ×35.
Lateral view.

☐ 3. UB 1238 (sample 918): Grönhögen, zone Vc; ×35.
Lateral view.

☐ 4. UB 1239 (sample 6420): Trolmen, zone Vc; ×50.
Posterolateral view.

☐ 5, 8. UB 1240 (sample 6223): Ödegården, zone Va?; ×50.
Lateral and basal view.

☐ 6. UB 1241 (sample 5947): Stenstorp–Dala, zone Vb; ×50.
Lateral view.

☐ 7. UB 1242 (sample 6482): Mörbylilla–Albrunna, zone V undiff.; ×50.
Oblique lateral view.

☐ 9. UB 1243 (sample 5650): Grönhögen, zone Vc; ×50.
Oblique posterior view.

☐ 10. UB 1244 (sample 926): Grönhögen, zone Vc; ×50.
Lateral view.

☐ 14. UB 1245 (sample 1000): Grönhögen, zone III; ×50.
Oblique basal view.

☐ 15. UB 1246 (sample 7230): S. Möckleby–Degerhamn, zone Vc; ×35.
Lateral view.

☐ 18. UB 1247 (sample 1000): Grönhögen, zone III; ×50.

Lateral view.

☐ 19, 20. UB 1248 (sample 5659): Degerhamn, zone Vc; ×50.
Lateral and basal view.

☐ 22. UB 1249 (sample 7231): Grönhögen, zone Vc; ×50.
Posterolateral view.

11–13, 16, 17. *Problematoconites angustus* n.sp., p. 36, all ×80.

☐ 11. UB 1250 (sample 5650): Trolmen, zone Vc.
Lateral view.

☐ 12, 13. UB 1251 (sample 7219): Grönhögen, zone Vc.
Holotype.
Lateral and basal view.

☐ 16. UB 1252 (sample 6345): Skår, zone V undiff.
Lateral view.

☐ 17. UB 1253 (sample 5650): Grönhögen, zone Vc.
Lateral view.

21, 23–26. *Problematoconites asymmetricus* n.sp., p. 36, all ×50.

☐ 21, 24. UB 1254 (sample 918): Grönhögen, zone Vc.
Holotype.
Posterior and basal view.

☐ 23. UB 1255 (sample 7232): S. Möckleby–Degerhamn, zone Vc.
Lateral view.

☐ 25. UB 1256 (sample 918): S. Möckleby–Degerhamn, zone Vc.
Posterolateral view.

☐ 26. UB 1505 (sample 918): Grönhögen, zone Vc.
Anterior view from above.

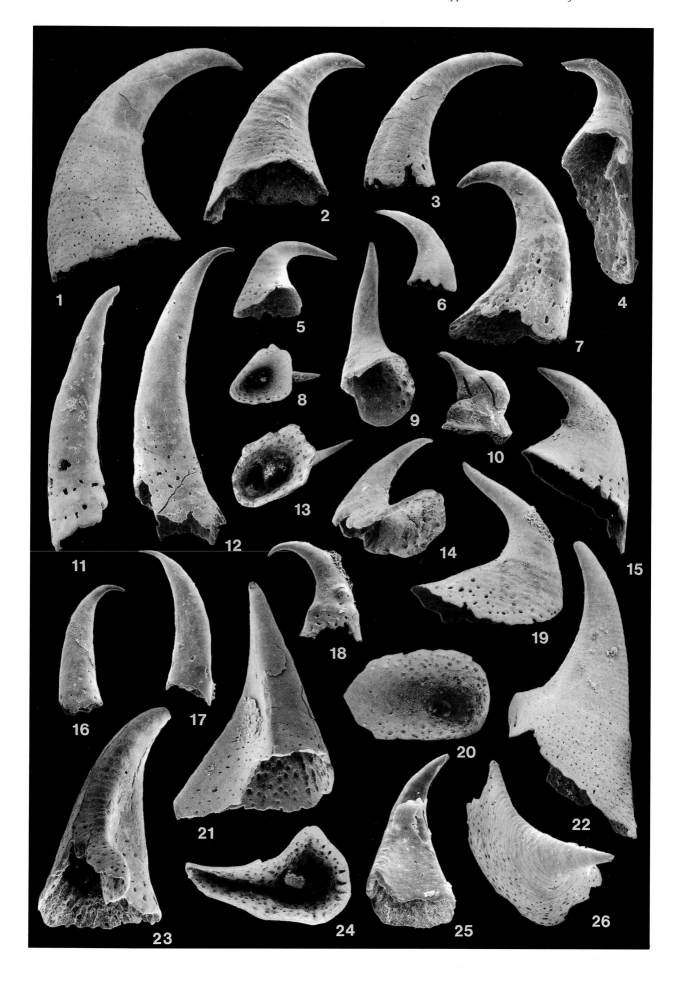

Plate 24

1–28. *Prooneotodus gallatini* (Müller 1959), p. 37, all ×120.

☐ 1, 5. UB 1257 (sample 7220): S. Möckleby–Degerhamn, zone Vc.
Lateral and basal view.

☐ 2, 6. UB 1258 (sample 7229): S. Möckleby–Degerhamn, zone Vc.
Lateral and basal view.

☐ 3, 7. UB 1259 (sample 6419): Trolmen, zone Vc.
Lateral and basal view.

☐ 4, 8. UB 1260 (sample 5965): Trolmen, zone V undiff.
Lateral and basal view.

☐ 9, 12. UB 1261 (sample 6210): Stenstorp–Dala, zone Vc.
Lateral and basal view.

☐ 10. UB 1262 (sample 7232): S. Möckleby–Degerhamn, zone Vc.
Cluster of three specimens, possibly coprolithic.

☐ 11, 16. UB 1263 (sample 6211): Stenstorp–Dala, zone Vb.
Lateral and basal view.

☐ 13. UB 1264 (sample 7219): Möckleby–Degerhamn, zone Vc.
Lateral view. Note the pronounced step between generally smooth upper and coarsely annulated lower part of the sclerite.

☐ 14, 17. UB 1265 (sample 6226): Ödegården, zone Va?.
Lateral and basal view.

☐ 15, 18. UB 1266 (sample 5967): Trolmen, zone V undiff.
Lateral and basal view.

☐ 19, 20. UB 1267 (sample 6268): Smedsgården–Stutagården, zone V undiff.
Lateral and basal view.

☐ 21, 22. UB 1268 (sample 5949): Stenstorp–Dala, zone Vb.
Lateral and basal view.

☐ 23, 24. UB 1269 (sample 6727): Trolmen, zone Vc.
Lateral and basal view.

☐ 25. UB 1270 (sample 6727): Trolmen, zone Vc.
Oblique lateral view of pathologically united sclerites ('siamese twins').

☐ 26. UB 1271 (sample 5964): Trolmen, zone Vc.
Posterolateral view of cluster with two sclerites.

☐ 27, 28. UB 1272 (sample 6419): Trolmen, zone Vc.
Lateral and basal view.

29. *Prooneotodus gallatini*? (Müller 1959).

☐ 29. UB 1273 (sample 5650): Grönhögen, zone Vc; ×70.
Anterior view from above. Very large specimen with almost circular cross-section and few boreholes close to the lower rim.

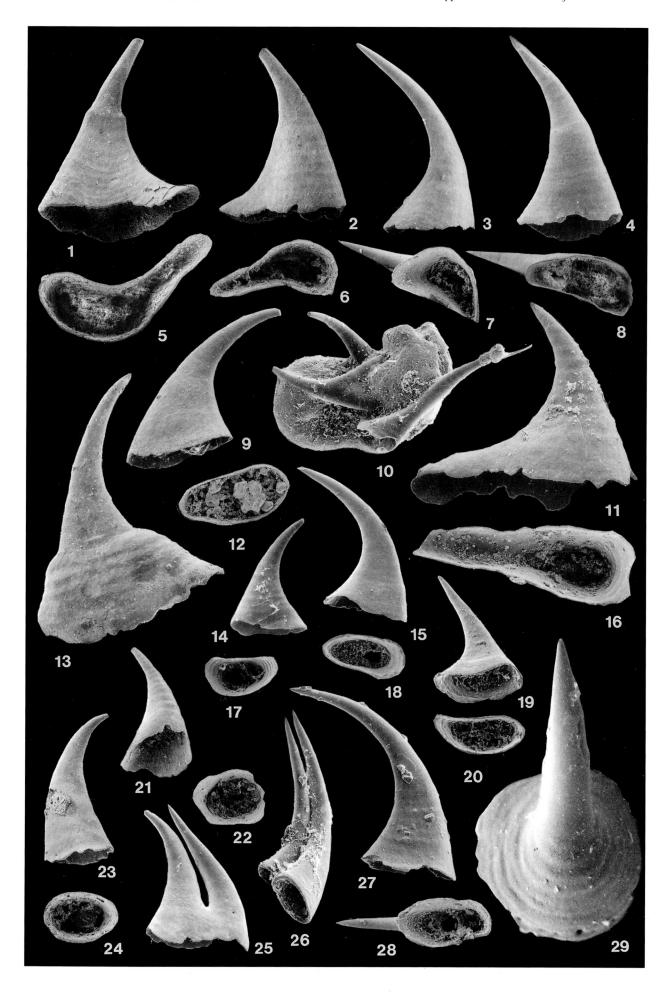

Plate 25

1–22. *Prosagittodontus dahlmani* Müller 1959, p. 37.

☐ 1, 2, 6. UB 1274 (sample 6406): Gum, zone Vc.
1, 2. Lateral and posterior view, ×70.
6. Basal view, ×80.

☐ 3, 4, 7. UB 1275 (sample 5659): Degerhamn, zone Vc.
3, 4. Posterior and lateral view, ×60.
7. Basal view, ×80.

☐ 5. UB 1276 (sample 5968): Trolmen, zone V undiff.; ×70.
Posterior view.

☐ 8, 12. UB 1277 (sample 6209): Stenstorp–Dala, zone Vc;
×70.
Posterior and oblique lateral view.

☐ 9. UB 1278 (sample 7217): S. Möckleby–Degerhamn,
zone Vc; ×60.
Posterior view.

☐ 10. UB 1279 (sample 5652): Grönhögen, zone Vc; ×70.
Posterior view.

☐ 11, 15. UB 1280 (sample 6406): Gum, zone Vc; ×70.

Anterior and lateral view.

☐ 13. UB 1281 (sample 6421): Trolmen, zone Vc; ×70.
Posterior view.

☐ 14, 19. UB 1282 (sample 6727): Trolmen, zone Vc; ×70.
Posterior and basal view of an asymmetrical, laterally
deflected specimen.

☐ 16. UB 1283 (sample 6793): Stubbegården, zone Vc; ×70.
Posterior view.

☐ 17. UB 1284 (sample 6337): Ekeberget, zone Vb; ×70.
Anterolateral view.

☐ 18. UB 1285 (sample 946): Karlsro, zone Vc; ×70.
Oblique posterior view.

☐ 20, 22. UB 1286 (sample 6406): Gum, zone Vc.
20. Posterior view, ×70.
22. Basal view, ×80.

☐ 21. UB 1287 (sample 6766): Gum, zone I; ×80.
Oblique anterior view of a short, but extremely expanded
specimen.

Plate 26

1–9, 11. *Trolmenia acies* n.gen., n.sp., p. 39.

☐ 1, 7. UB 1288 (sample 5970): Trolmen, zone Vc.
   1. Posterolateral view, ×70.
   7. Basal view, ×90.

☐ 2, 8. UB 1289 (sample 5650): Grönhögen, zone Vc.
   2. Posterolateral view, ×90.
   8. Basal view, ×120.

☐ 3. UB 1290 (sample 6186): Stenstorp–Dala, zone Vb; ×90.
   Oblique lateral view.

☐ 4, 9. UB 1291 (sample 6238): Ödegården, zone Vb; ×90.
   Holotype.
   Lateral and basal view.

☐ 5, 6. UB 1292 (sample 5650): Grönhögen, zone Vc; ×90.
   Lateral and anterior view.

☐ 11. UB 1293 (sample 6421): Trolmen, zone Vc; ×90.
   Cluster of three specimens.

10, 12–24. *Prosagittodontus minimus* n.sp., p. 38.

☐ 10, 16. UB 1294 (sample 6162): Stenstorp–Dala, zone Vc.
   10. Posterior view, ×120.
   16. Basal view, ×150.

☐ 12, 13. UB 1295 (sample 7225): S. Möckleby–Degerhamn, zone Vc; ×120. Holotype.
   Posterior and lateral view.

☐ 14, 15. UB 1296 (sample 6230): Ödegården, zone Va?; ×120.
   Posterior and basal view.

☐ 17. UB 1297 (sample 926): Grönhögen, zone Vc; ×120.
   Posterior view.

☐ 18. UB 1298 (sample 7225): S. Möckleby–Degerhamn, zone Vc;×120.
   Lateral view.

☐ 19, 23. UB 1299 (sample 6164): Stenstorp–Dala, zone Vc; ×120.
   Posterior and basal view.

☐ 20. UB 1300 (sample 6343): Skår, zone Vc; ×120.
   Lateral view.

☐ 21, 24. UB 1301 (sample 6406): Gum, zone Vc; ×120.
   Posterior and basal view.

☐ 22. UB 1302 (sample no 5659): Degerhamn, zone Vc; ×120.
   Posterior view.

Plate 27

1–17. *Serratocambria minuta* n.gen., n.sp., p. 38, all ×200.

☐ 1, 3. UB 1303 (sample 5951): Trolmen, zone Vc.
Posterior and basal view.

☐ 2, 4, 7. UB 1304 (sample 5951): Stenstorp–Dala, zone Vc.
Posterior, basal, and lateral view.

☐ 5, 6. UB 1305 (sample 6190): Stenstorp–Dala, zone Vc.
Posterior and basal view.

☐ 8, 9. UB 1306 (sample 5948): Stenstorp–Dala, zone Vc.
Posterior and basal view.

☐ 10, 11. UB 1307 (sample 6228): Ödegården, zone Vb?.
Holotype.
Posterior and basal view.

☐ 12, 13. UB 1308 (sample 6407): Gum, zone Vc.
Anterior view from above and oblique posterior view.

☐ 14, 17. UB 1309 (sample 5947): Stenstorp–Dala, zone Vb.
View from above and oblique anterior view.

☐ 15, 16. UB 1310 (sample 5932): Stenstorp–Dala, zone Vc.
Posterior and basal view.

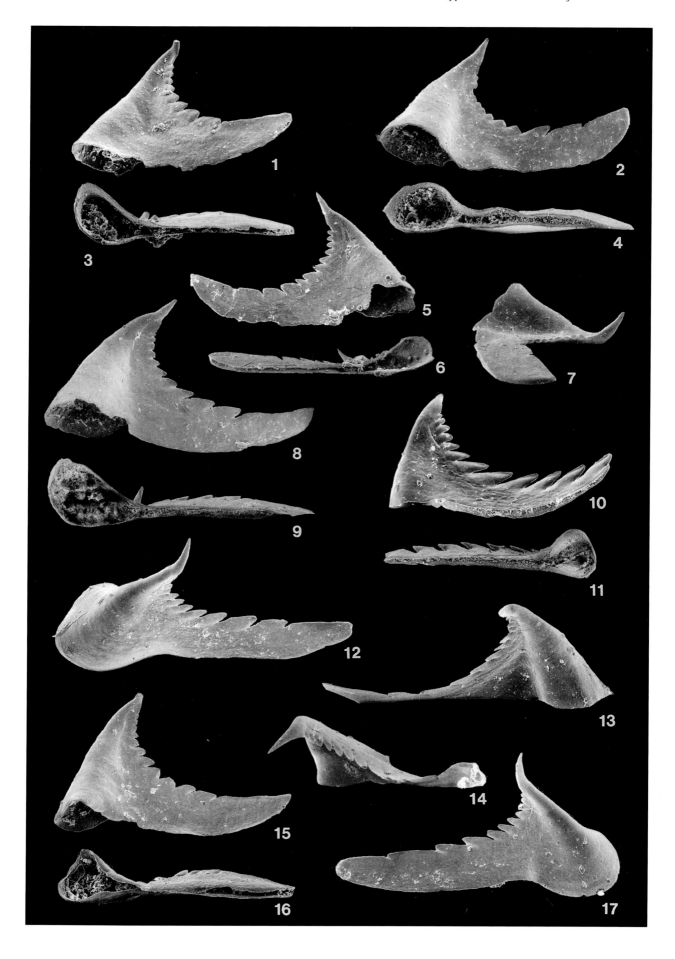

Plate 28

1–14. *Westergaardodina ligula* n.sp., p. 46, all ×140.

☐ 1, 7, 8. UB 1311 (sample 6172): Stenstorp–Dala, zone Vb. Holotype.
Posterior, basal, and lateral view. Note the deeply excavated posterior side with marginal bulge and tiny lateral openings.

☐ 2, 3. UB 1312 (sample 6388): Haggården–Marieberg, zone Vc.
2. Oblique view onto the posterior side photographed from the tips which involved perspective shortening of the lateral projections.
3. Posterior view.

☐ 4, 5. UB 1313 (sample 6238): Ödegården, zone Vb or c.
Anterior view with spine–like median projection and profile.

☐ 6. UB 1314 (sample 6710): Ödegården, zone Vb.
Anterior view.

☐ 9, 10. UB 1315 (sample 6238): Ödegården, zone Vb.
Lateral and posterior view.

☐ 11, 12. UB 1316 (sample 6185): Stenstorp–Dala, zone Vc.
Lateral and posterior view.

☐ 13, 14. UB 1317 (sample 6369): Brattefors, zone V undiff.
Profile and posterior view of mature specimen with accordingly enlarged lateral openings.

15–20. *Westergaardodina matsushitai* Nogami 1966, p. 46.

☐ 15, 16. UB 1318 (sample 7214): S. Möckleby–Degerhamn, zone V undiff.; ×40.

Posterior and oblique lateral view, showing extremely long lateral openings. The basis itself is closed.

☐ 17. UB 1319 (sample 6404): Haggården–Marieberg, zone II; ×30.
Posterior side. Broken basal portion permits a view into the hollow inner side with outcropping growth lamellae.

☐ 18. UB 1320 (sample 6735): Backeborg, zone I; ×50.
Lateral view.

☐ 19. UB 1321 (sample 6768): Gum, zone I; ×50.
Anterior view.

☐ 20. UB 1322 (sample 6735): Backeborg, zone I; ×50.

21–27. *Westergaardodina wimani* Szaniawski 1971, p. 51, all ×80.

☐ 21, 27. UB 1323 (sample 6739): Klippan, zone I.
Lateral and anterior view showing the flattened anterior face with median callosity.

☐ 22. UB 1324 (sample 6776): Gum, zone I.
Anterior view of large specimen with circular notch between the lateral projections. It probably marks the position of a phosphatic ball. Such spheres are often attached to the lateral projections of bicuspidate elements.

☐ 23–25. UB 1325 (sample 6776): Gum, zone I.
Posterior and lateral views demonstrating that the shorter lateral projection is closed, the other one marked by a shallow opening.

☐ 26. UB 1326 (sample 6776): Gum, zone I.
Posterior view.

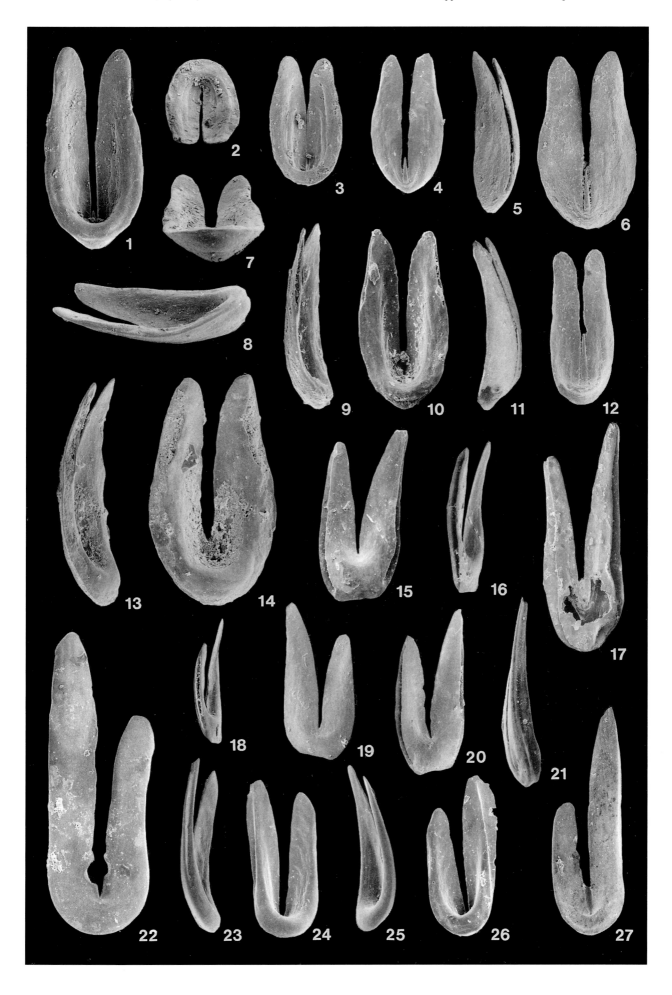

Plate 29

1–12. *Westergaardodina bohlini* Müller 1959, p. 43.

☐ 1, 2. UB 1327 (sample 5662): Degerhamn, zone Vc; ×110. Posterior and lateral view with stout, keeled median projection and elongate lateral openings.

☐ 3, 4. UB 1328 (sample 5962): Trolmen, zone Vc; ×125. Anterior and lateral view of a very small specimen. The lateral projections are in touch with each other above the broken median projection.

☐ 5. UB 1329 (sample 5662): Degerhamn, zone Vc; ×125. Anterior view of specimen of almost rectangular outline with well-developed callosities around the turning points and comarginal folds to the basis.

☐ 6, 8. UB 1330 (sample 6710): Ödegården, zone Vb or c; ×125.
Lateral and anterior view.

☐ 7. UB 1331 (sample 6476): Degerhamn, zone IV or Va. View from the basis onto the anterior side with pronounced callosities.

☐ 9. UB 1332 (sample 6230): Ödegården, zone Va?; ×110. Posterior view.

☐ 10. UB 1333 (sample 6238): Ödegården, zone Vb; ×125.

Posterior view.

☐ 11. UB 1334 (sample 7559): Karlsfors, zone IV; ×80. Anterior view of specimen with flaring lateral projections and straight, closed top ends.

☐ 12. UB 1335 (sample 6429): Karlsfors, zone IV; ×80. A similar specimen from the posterior side.

13–19. *Westergaardodina bicuspidata* Müller 1959, p. 42.

☐ 13. UB 1336 (sample 6313): Tomten, zone Vc; ×60. Anterior view.

☐ 14. UB 1337 (sample 695) Gallatin Limestone, Port Clear Creek, W Buffalo, Wyoming, USA; ×60.
Anterior view.

☐ 15, 19. UB 1338 (sample 6320): Tomten, zone Vc; ×60. Basal and posterior view.

☐ 16. UB 1339 (sample 6142): Stenstorp–Dala, zone Vb; ×40.
Posterior view.

☐ 17, 18. UB 1340 (sample 6220): Ödegården, zone Vc; ×60.
Lateral and posterior view.

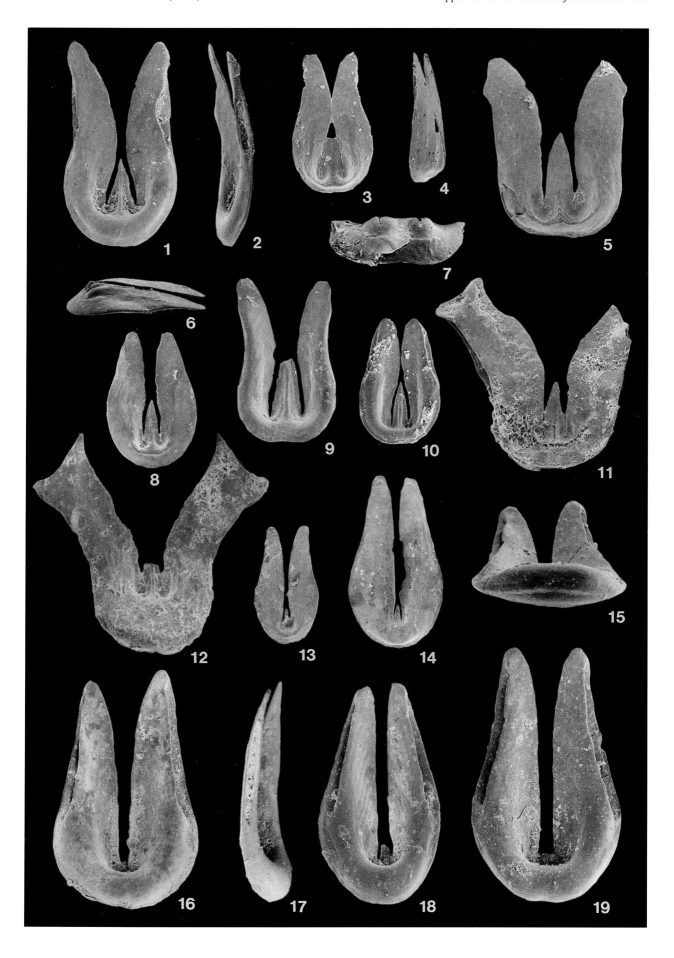

Plate 30

1–8, 10. *Westergaardodina moessebergensis* Müller 1959, p. 47.

☐ 1, 6. UB 1341 (sample 6739): Klippan, zone I; ×60. Posterior and side view showing large continuous cavity. The posterior side is distinctly smaller than the anterior one.

☐ 2. UB 1342 (sample 6763): Gum, zone I; ×60. Anterior view.

☐ 3. UB 1343 (sample 994): Sätra, zone I; ×80. Anterior view.

☐ 4. UB 1344 (6743): Klippan, zone I; ×80. Posterior view.

☐ 5. UB 1345 (sample 6746): St. Stolan, zone I; ×120. Posterior view.

☐ 7. UB 1346 (sample 6779): Gum, zone I; ×60. Basal view. The centrally bridged anterior and posterior faces reduce the basal opening to a shallow furrow.

☐ 8. UB 1347 (sample 6746): St. Stolan, zone I; ×120. Posterior view of very small specimen. These stages have a wider bridge along the basal margin than larger elements.

☐ 10. UB 1348 (sample 6735): Backeborg, zone I; ×60. Anterior view.

9, 11–21. *Westergaardodina quadrata* (An 1982), p. 50.

☐ 9. UB 1349 (sample 6761): Gum, zone I; ×50. Anterior view.

☐ 11. UB 1350 (sample 6417): Gum, zone I; ×70. Anterior view.

☐ 12. UB 1351 (sample 6414): Gum, zone I; ×100. Lateral view.

☐ 13. UB 1352 (sample 6731): Backeborg, zone I; ×50. Oblique basal view showing a faint undulation of the anterior side.

☐ 14. UB 1353 (sample 6763): Gum, zone I; ×50. Upper posterior view of very large specimen with broken lateral projections. Note the bowl-shaped posterior side with only little contact to the anterior one.

☐ 15. UB 1354 (sample 6412): Gum, zone I; ×100. Posterior view of small specimen with both anterior and posterior side tightly attached to each other along the basal rim.

☐ 16. UB 1355 (sample 6739): Kakeled, zone I; ×100. Anterior view.

☐ 17. UB 1356 (sample 6735): Backeborg, zone I; ×50. Oblique posterior view. The pits covering the entire surface of the basal opening are probably due to parasitism.

☐ 18. UB 1357 (sample 6787): Gum, zone I; ×50. Posterior view.

☐ 19. UB 1358 (sample 6764): Gum, zone I; ×70. Basal view of intermediate stage with still fairly wide bridge at the basis.

☐ 20, 21. UB 1359 (sample 6784): Gum, zone I; ×50. Profile and posterior view.

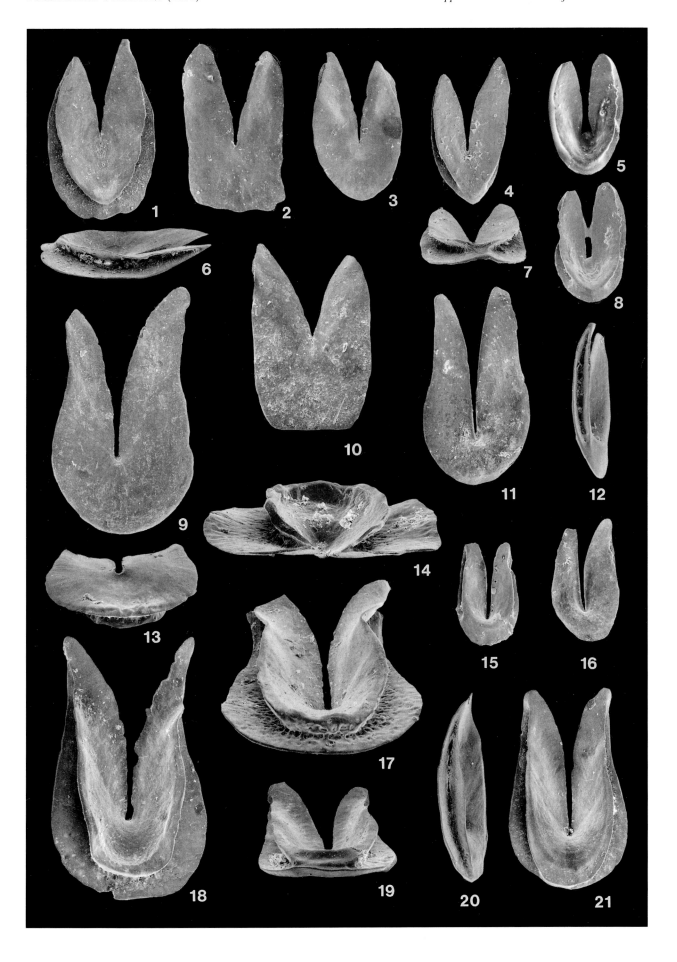

Plate 31

1–21. *Westergaardodina polymorpha* n.sp., p. 48.

☐ 1, 2. UB 1360 (sample 6323): Tomten, zone Vb; ×70.
Posterior view and profile.

☐ 3, 4, 7. UB 1361 (sample 6185): Stenstorp–Dala, zone Vc;
×70.
Posterior, oblique, and basal view showing distinct furrow connecting the lateral openings.

☐ 5, 6. UB 1362 (sample 7217): S. Möckleby–Degerhamn,
zone Vc; ×70.
Anterior side with small but distinct median projection
and profile.

☐ 8. UB 1363 (sample 7232): S. Möckleby–Degerhamn,
zone Vc; ×40.
Anterior side of larger specimen with much increased
posterior side in the basal portion. Note the growth
lamellae.

☐ 9. UB 1364 (sample 6425): Karlsfors, zone Vc; ×40.
Anterior view of fragment of a large specimen with scalloped posterior basal margin.

☐ 10. UB 1365 (sample 5964): Trolmen, zone Vb; ×70.
Posterior side with large phosphatic ball.

☐ 11, 12. UB 1366 (sample 7217): S. Möckleby–Degerhamn, zone Vc; ×40.
Posterior and posterolateral view.

☐ 13. UB 1367 (sample 5673): S. Möckleby–Degerhamn,
zone Vb?; ×40. Holotype.

Anterior view with tiny median projection. The posterior
side is circularly enlarged in the basal part.

☐ 14. UB 1368 (sample 7217): S. Möckleby–Degerhamn,
zone Vc; ×70.
Basal view , showing growth lamellae along the inner
margin of the lateral projections.

☐ 15. UB 1369 (sample 7217): S. Möckleby–Degerhamn,
zone Vc; ×40.
Anterolateral view.

☐ 16. UB 1370 (sample 6238): Ödegården, zone Vb; ×70.

☐ 17. UB 1371 (sample 6164): Stenstorp–Dala, zone Vc;
×70.

☐ 18. UB 1372 (sample 6296): Ranstadsverket, zone Vc;
×40.
Anterior side with circular notch between the lateral
projections, probably from a phosphatic ball during
early diagenetic compaction.

☐ 19. UB 1373 (sample 6388): Haggården–Marieberg,
zone Vc; ×40.
Anterior view of big, fragmentary specimen with much
enlarged posterior side.

☐ 20. UB 1374 (sample 6710): Stenstorp–Dala, zone Vb or
c; ×70.
Anterior view.

☐ 21. UB 1375 (sample 6710): Stenstorp–Dala, zone Vb or
c; ×40.
Anterolateral view.

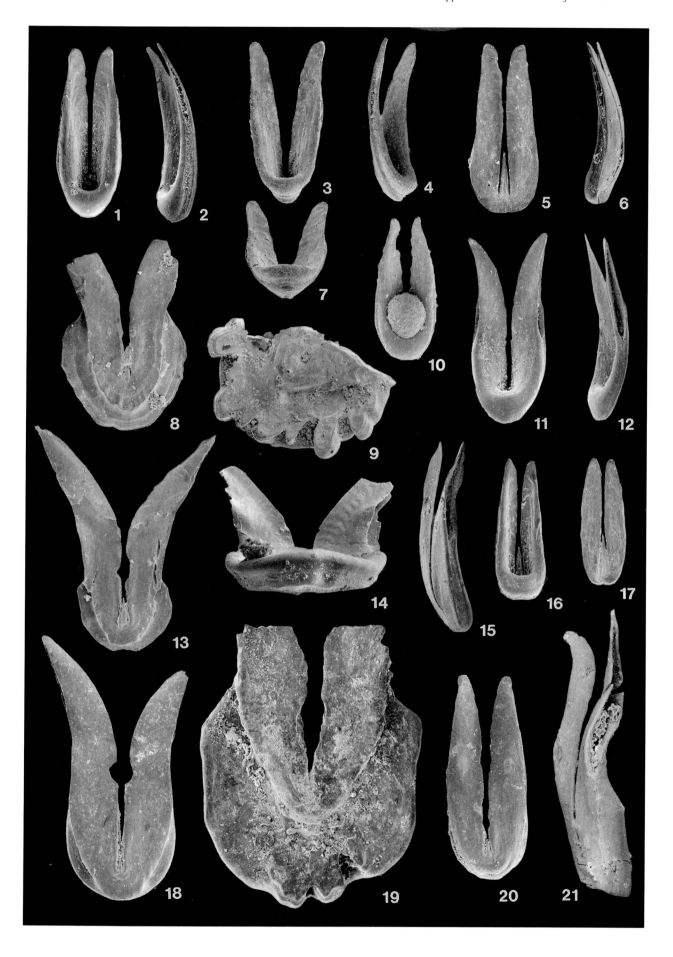

Plate 32

1–18. *Westergaardodina procera* n.sp., p. 49.

☐ 1, 2, 6. UB 1376 (sample 5650): Grönhögen, zone Vc; ×110. Holotype.
Oblique basal, posterior, and lateral view. Note the lateral projections which overlap each other beyond the comparatively large median projection.

☐ 3. UB 1377 (sample 926): Ranstadsverket, zone Vc; ×110. Anterior view.

☐ 4. UB 1378 (sample 5672): Grönhögen, zone Vc; ×90. Posterior view.

☐ 5. UB 1379 (sample 6301): Rörsberga, zone Vc; ×90. Anterior view.

☐ 7. UB 1380 (sample 6153): Grönhögen, zone Vc; ×110. Posterior view.

☐ 8. UB 1381 (sample 5948): Stenstorp–Dala, zone Vb; ×110.
Oblique lateral view of small specimen. Note punctual breakage documenting that the bulge is hollow.

☐ 9. UB 1382 (sample 926): Degerhamn, zone Vc; ×160. Basal view.

☐ 10. UB 1383 (sample 6294): Stenstorp–Dala, zone Vb; ×110.
Oblique anterior view with angulate face.

☐ 11. UB 1384 (sample 5650): Stenstorp–Dala, zone Vb; ×110.
Anterolateral view.

☐ 12. UB 1385 (sample 5659): Grönhögen, zone Vc; ×90. Anterior view.

☐ 13. UB 1386 (sample 5948): S. Möckleby, zone Vb; ×90. Posterior view of specimen with attached phosphatic ball.

☐ 14. UB 1387 (sample 926): Grönhögen, zone Vc; ×90. Posterior view of specimen with large ball.

☐ 15. UB 1388 (sample 918): Grönhögen, zone Vc; ×90. Posterior view.

☐ 16, 17. UB 1389 (sample 990): Grönhögen, zone Vc; ×90. Lateral and anterior view. Anterior side shows longitudinal furrow on the median process extending downwards to the marginal rim.

☐ 18. UB 1390 (sample 926): Trolmen, zone Vc; ×90. Posterior view.

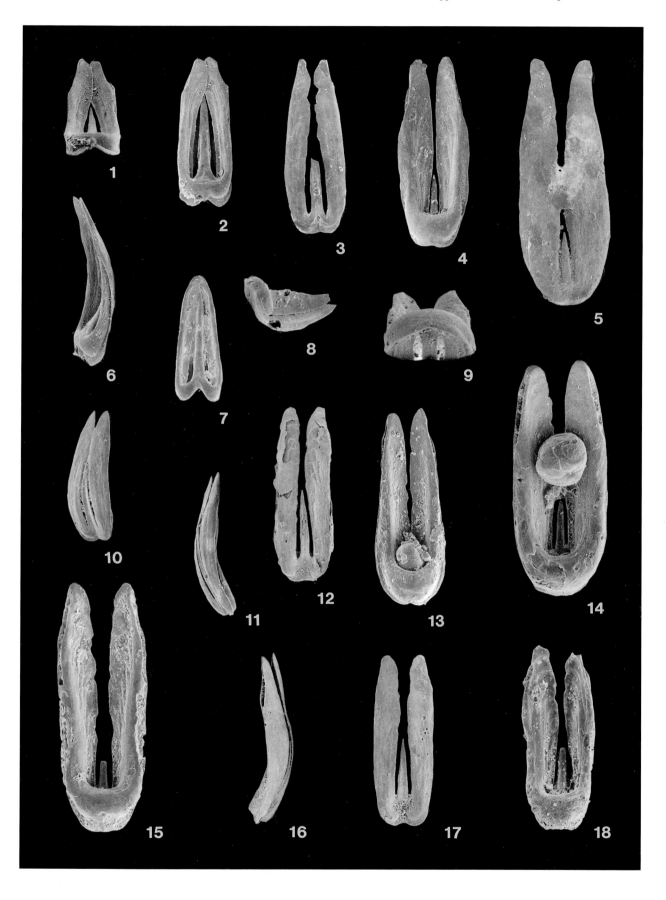

Plate 33

1, 2, 4. *Westergaardodina obliqua* Szaniawski 1971, p. 48.

☐ 1, 2, 4. UB 1391 (sample 6409): Gum, zone I; ×120.
Posterior and both lateral views.

3, 5–16. *Westergaardodina excentrica* n.sp., p. 45, all ×120.

☐ 3, 5, 8. UB 1392 (sample 6417): Gum, zone I. Holotype.
Upper, posterior and posterior–oblique basal view. Median keel does not coincide with the direction of the median projection.

☐ 6, 7. UB 1393 (sample 6761): Gum, zone I.
Lateral and anterior view. Note the growth lamellae.

☐ 9. UB 1394 (sample 994): Kakeled, zone I.

☐ 10, 12. UB 1395 (sample 6409): Gum, zone I.
Anterior and anterior–oblique basal view.

☐ 11. UB 1396 (sample 6363): St. Stolan, zone I.
Upper view onto the posterior side.

☐ 13, 15, 16. UB 1397 (sample 5952): Stenstorp–Dala, zone V undiff.

Oblique anterior views of a small, extremely twisted specimen.

☐ 14. UB 1398 (sample 6730): Backeborg, zone I.
Anterior view.

17–22. *Westergaardodina amplicava* Müller 1959, p. 41.

☐ 17. UB 1399 (sample 5650): Grönhögen, zone Vc; ×50.
Posterior view.

☐ 18. UB 1400 (sample 6313): Tomten, zone Vb; ×30.
Oblique posterior view.

☐ 19, 20. UB 1401 (sample 5650): Grönhögen, zone Vc; ×50.
Anterior view from above and lateral view. Partly exfoliated coating reveals slight callosities around the turning points and faint growth lamellae.

☐ 21. UB 1402 (sample 6338): Ekeberget, zone Vc; ×30.
Posterior view.

☐ 22. UB 1403 (sample 7224): S. Möckleby–Degerhamn, zone Vc; ×30.
Posterior view.

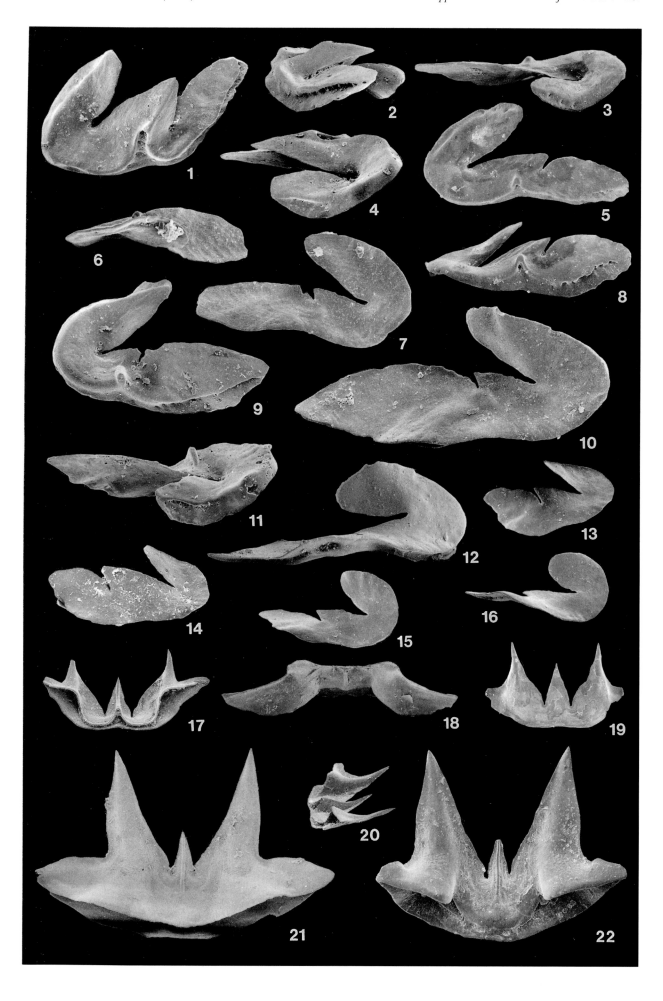

Plate 34

1, 2, 4. *Westergaardodina amplicava* Müller 1959, p. 41.

☐ 1. UB 1404 (sample 926): Grönhögen, zone Vc; ×50.
Anterior view. Due to preservation, the posterior side extends much beyond the anterior one and has growth lines exposed on its inner surface.

☐ 2. UB 1405 (sample 6234): Ödegården, zone Vb; ×30.
Anterior side with well–developed, almost rectangularly spread lateral processes.

☐ 4. UB 1406 (sample 7223): S. Möckleby–Degerhamn, zone V undiff.; ×50.
Posterior side; lateral processes of the lateral projections are not preserved.

3, 5–12, 15, 16. *Westergaardodina concamerata* n.sp., p. 44.

☐ 3, 7, 8. UB 1407 (sample 5650): Grönhögen, zone Vc; ×160.
Oblique basal, posterior and lateral view. Flat specimen without basal opening.

☐ 5, 11, 12. UB 1408 (sample 5970): Trolmen, zone Vc; ×160. Holotype.
Oblique posterior, posterior, and posterolateral view. Note callosities around the turning points and the hollow keel which continues in a sheet.

☐ 6. UB 1409 (sample 5962): Trolmen, zone Vc; ×160.

Oblique side view, documenting the extremely flat shape.

☐ 9, 10. UB 1410 (sample 5962): Trolmen, zone Vc; ×160.
Oblique posterior and posterolateral view.

☐ 15. UB 1411 (sample 5970): Trolmen, zone Vc; ×160.
Anterior view.

☐ 16. UB 1412 (sample 926): Grönhögen, zone Vc; ×250.
Anterior view of tiny specimen with widely arched basal margin. Note the exfoliated coating.

13, 14, 17–21. *Westergaardodina auris* n.sp., p. 41, all ×40.

☐ 13, 14, 17. UB 1413 (sample 6411): Gum, zone I. Holotype.
Oblique lateral, basal, and posterior view. Note solid keel and bulgy posterior margin. The latter has ear-like widenings on the lateral projections.

☐ 18, 21. UB 1414 (sample 6411): Gum, zone I.
Lateral and posterior view. The continuous basal opening widens along the lateral projections to well-developed ears.

☐ 19. UB 1415 (sample 6784): Gum, zone I.
Anterior side with crenulate basal rim and largely closed top ends of the lateral projections.

☐ 20. UB 1416 (sample 6783): Gum, zone I.
Basal view.

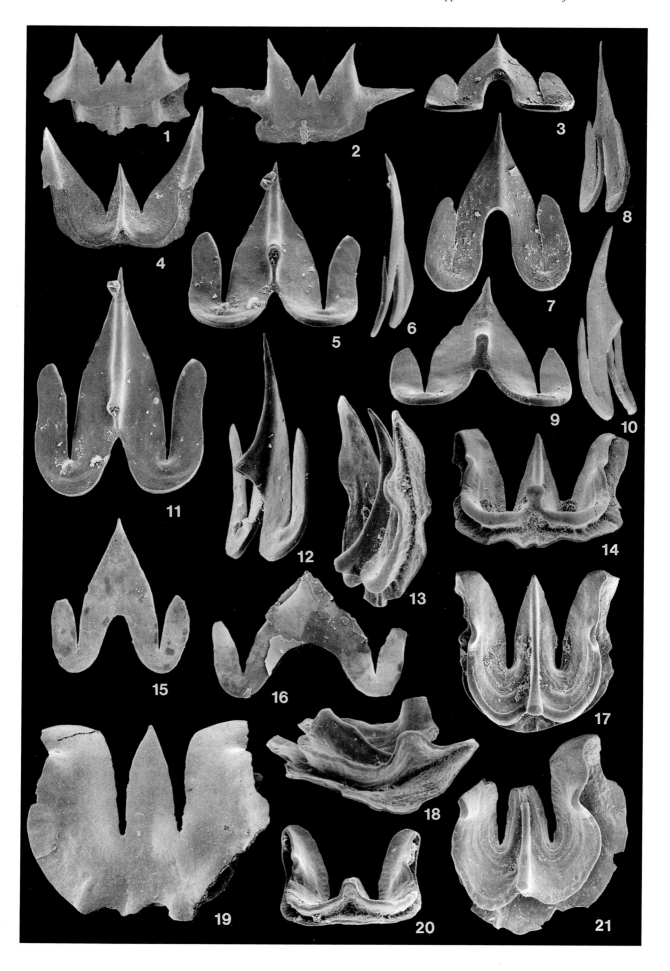

Plate 35

1–9, 12: *Westergaardodina curvata* n.sp., p. 45, all ×130.

☐ 1. UB 1417 (sample 6727): Trolmen, zone Vc.
Oblique posterior view.

☐ 2, 5, 8. UB 1418 (sample 926): Grönhögen, zone Vc.
Holotype.
Basal, lateral, and posterior view. Note the flat, slightly convex shape with pronounced, hollow, median keel. Distinct callosities on the posterior side.

☐ 3, 6, 9. UB 1419 (sample 926): Grönhögen, zone Vc.
Posterior, oblique posterior, and lateral view.

☐ 4, 7. UB 1420 (sample 926): Grönhögen, zone Vc.
Posterior and upper views. Note recurvature of median and lateral tips.

☐ 12. UB 1421 (sample 5650): Grönhögen, zone Vc.
Anterior side of large specimen.

10, 11, 13–17: *Westergaardodina latidentata* n.sp., p. 46, all ×190.

☐ 10, 11. UB 1422 (sample 6211): Stenstorp–Dala, zone Vb.
Holotype.
Lateral and posterior views.

☐ 13. UB 1423 (sample 6302): Rörsberga, zone Vb.
Posterior side with incipient keel.

☐ 14. UB 1424 (sample 6357): St. Stolan, zone Vc.
Posterior view.

☐ 15, 16. UB 1425 (sample 6188): Stenstorp–Dala, zone Vc.
Oblique lateral and anterior views, the latter showing callosities around the deeply incised turning points.

☐ 17. UB 1426 (sample 6711): Stenstorp–Dala, zone Vc.
Posterior view of large specimen with fairly well-developed keel.

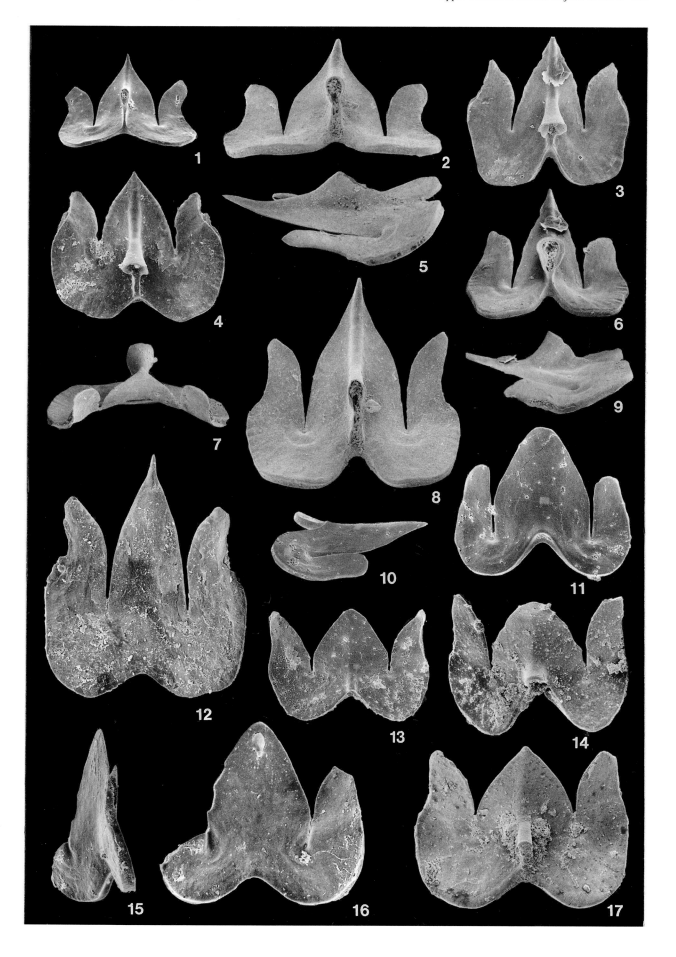

Plate 36

1–4, 6: *Westergaardodina tricuspidata* Müller 1959, p. 50, all ×75.

☐ 1, 2. UB 1427 (sample 6808): Toreborg, zone II.
Posterior and lateral views. Note the hole (parasitic?) on the bulgy marginal rim which is evidence of the hollow interior.

☐ 3, 6. UB 1428 (sample 922): Grönhögen, zone III.
Anterior and side view showing the deeply incised turning points with fold-like callosities around and the longitudinal depression on the median projection.

☐ 4. UB 1429 (sample 1000): Grönhögen, zone III.
Posterior view.

5, 7–17: *Westergaardodina communis* n.sp., p. 44.

☐ 5, 8, 9. UB 1430 (sample 6776): Gum, zone I; ×60.
Upper, oblique lateral and posterior views. Note the short, spur-like median keel and the broken, massive top ends of the lateral projections.

☐ 7. UB 1431 (sample 6736): Backeborg, zone I; ×60.
Oblique posterior view.

☐ 10, 13. UB 1432 (sample 6783): Gum, zone I; ×60. Holotype.
Posterior and lateral view.

☐ 11. UB 1433 (sample 6735): Backeborg, zone I; ×60.
Anterior side of a small specimen.

☐ 12. UB 1434 (sample 6748): Gum, zone I; ×55.
Side view with outcropping growth lamellae within the lateral openings.

☐ 14. UB 1435 (sample 6735): Backeborg, zone I; ×50.
Lateral view.

☐ 15. UB 1436 (sample 6735): Backeborg, zone I; ×60.
Posterior view.

☐ 16. UB 1437 (sample 6410): Gum, zone I; ×60.
Posterior view.

☐ 17. UB 1438 (sample 6787): Gum, zone I; ×55.
Posterior view of largest specimen, with distinctly outward directed tips of lateral projections.

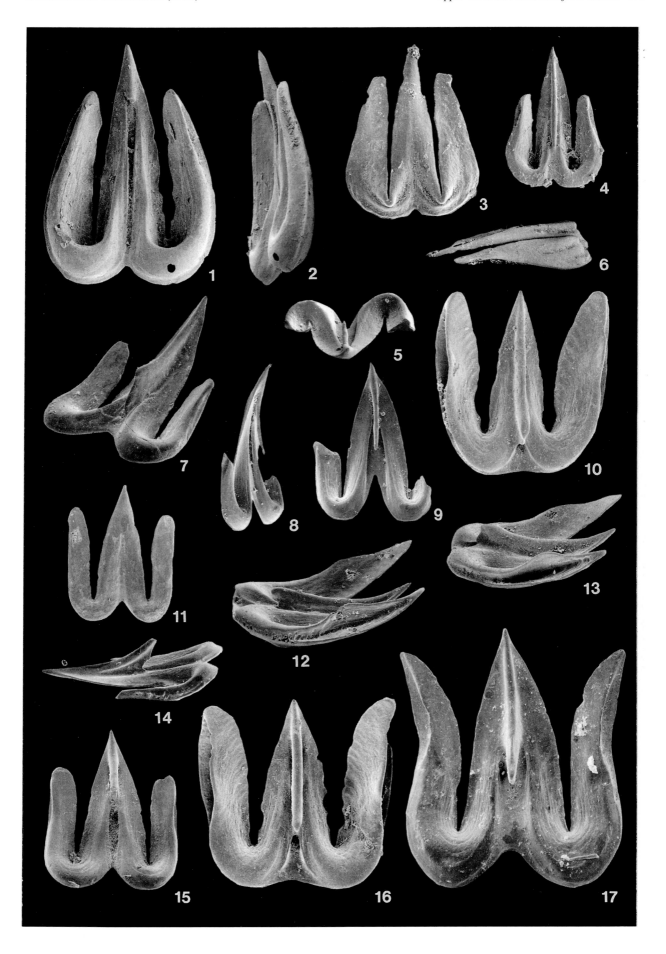

Plate 37

1–6, 9, 10, 12, 13: *Westergaardodina behrae* n.sp., p. 42, all ×160.

☐ 1, 2, 5. UB 1439 (sample 5942): Stenstorp–Dala, zone Vb.
Posterolateral, posterior, and basal view.

☐ 3, 4, 6. UB 1440 (sample 6186): Stenstorp–Dala, zone Vb.
Holotype.
Oblique lateral, posterior and basal views. Note the deeply excavated area around the turning points and the rod-like keel with sheet.

☐ 9. UB 1441 (sample 6238): Ödegården, zone Vb.
Oblique anterior view.

☐ 10. UB 1442 (sample 6211): Stenstorp–Dala, zone Vb.
Posterior view.

☐ 12. UB 1443 (sample 6238): Ödegården, zone Vb.
Anterior side with gentle, ring-like callosities.

☐ 13. UB 1444 (sample 6172): Stenstorp–Dala, zone Vb.

Posterior view.

7, 8, 11, 14–16: *Westergaardodina calix* n.sp., p. 43, all ×90.

☐ 7, 8. UB 1445 (sample 7232): S. Möckleby–Degerhamn, zone Vc. Holotype.
Oblique lateral and posterior views with comparatively low, rod-like keel.

☐ 11. UB 1446 (sample 6238): Ödegården, zone Vb.
Oblique anterior view.

☐ 14. UB 1447 (sample 6369): Brattefors, zone V undiff.
Anterior view. Note the comparatively strong lateral projections.

☐ 15. UB 1448 (sample 6369): Brattefors, zone V undiff.
Posterior view.

☐ 16. UB 1449 (sample 6369): Brattefors, zone V undiff.
Posterior side with well-developed lateral openings.

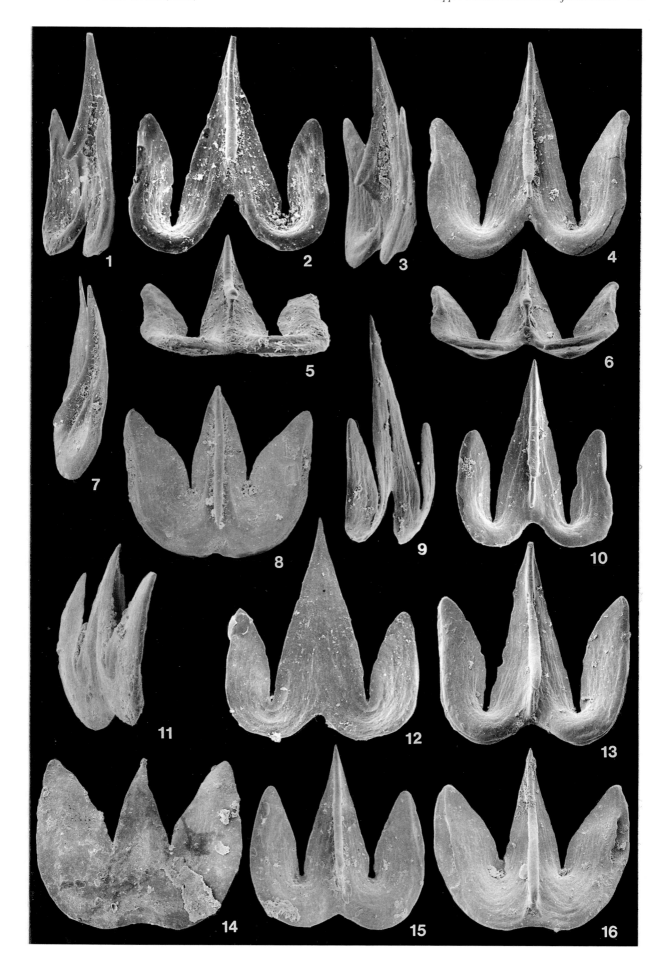

Plate 38

1–11. *Westergaardodina ahlbergi* n.sp., p. 41, all ×90.

☐ 1–3. UB 1450 (sample 6785): Gum, zone I. Holotype. Basal, posterior and side view showing the short, strongly projecting hollow keel and comarginal bulgings at the basis.

☐ 4, 6. UB 1451 (sample 6735): Backeborg, zone I. Posterior and basal view. Inside the hollow keel, traces of parasitic activity are visible.

☐ 5. UB 1452 (sample 6787): Gum, zone I. Anterior view.

☐ 7. UB 1453 (sample 6404): Haggården–Marieberg, zone II.

Posterior view.

☐ 8. UB 1454 (sample 6764): Gum, zone I. Posterior view.

☐ 9. UB 1455 (sample 6750): Gum, zone I. Posterolateral view.

☐ 10. UB 1456 (sample 6761): Gum, zone I. Posterior view.

☐ 11. UB 1457 (sample 6765): Gum, zone I. Anterior view of specimen with extremely diverging lateral projections.

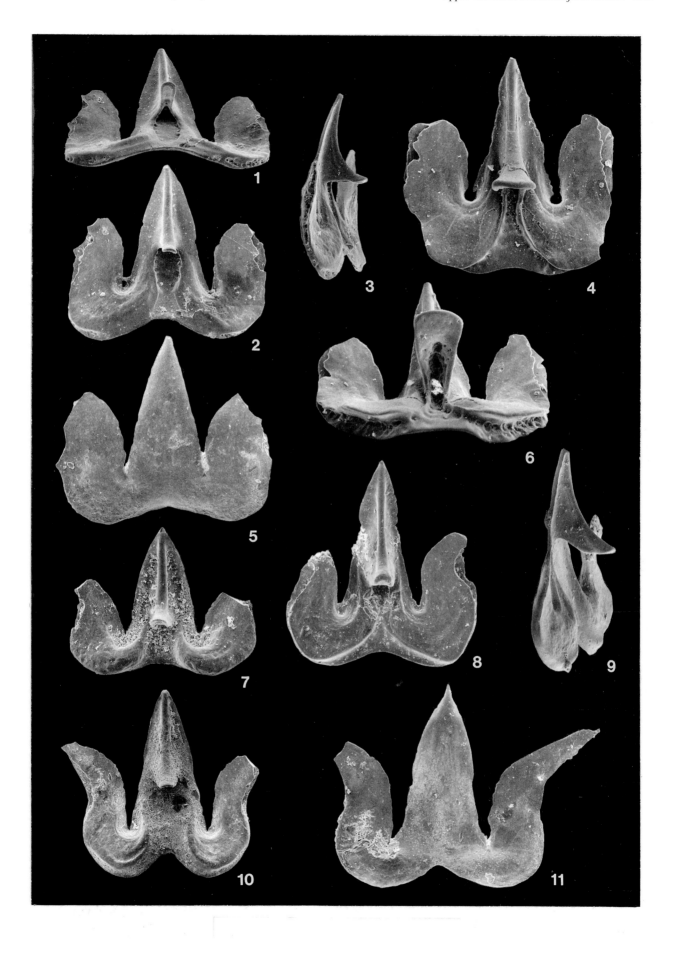

Plate 39

1–10. *Westergaardodina nogamii* n.sp., p. 47, all ×40.

☐ 1, 3. UB 1458 (sample 6735): Backeborg, zone I.
1. Basal view with large, hollow keel and continuous basal opening. Within the latter, numerous boreholes are visible.
3. Posterior side with median keel, which widens adapically into a rostrum.

☐ 2, 4, 5. UB 1459 (sample 6764): Gum, zone I. Holotype. Basal, posterior and lateral view showing the strongly projecting keel and inwardly curved lateral projections.

☐ 6, 9. UB 1460 (sample 6787): Gum, zone I.
Anterior side and oblique posterior view from above.

☐ 7. UB 1461 (sample 6750): Gum, zone I.
Anterior view.

☐ 8. UB 1462 (sample 6404): Haggården–Marieberg, zone II.
Anterior view of single specimen with indented lateral projections.

☐ 10. UB 1463 (sample 6762): Gum, zone I.
Posterior view of comparatively tall sclerite with narrow but still characteristic lateral projections.

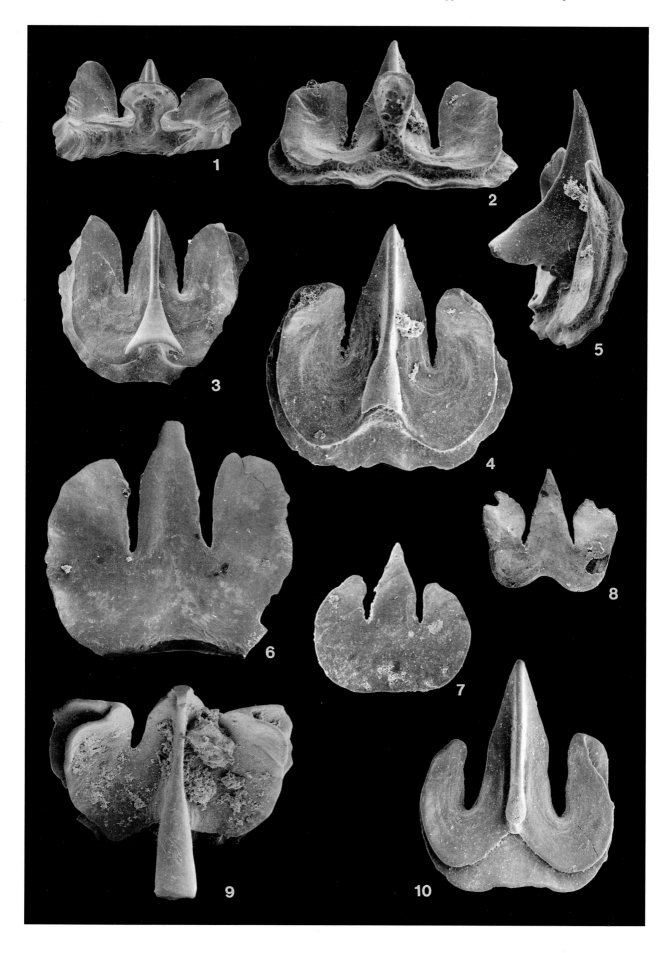

Plate 40

1–7: *Westergaardodina prominens* n.sp., p. 49, all ×210.

☐ 1, 4. UB 1464 (sample 6164): Stenstorp–Dala, zone Vc.
Posterior–basal and posterior view.

☐ 2, 3. UB 1465 (sample 6264): Smedsgården–Stutagården, zone Vb or c. Holotype.
Anterior and oblique lateral view. Note the extremely high, almost spherical callosities.

☐ 5, 7. UB 1467 (sample 6264): Smedsgården–Stutagården, zone Vb or c.
Posterior and lateral view.

☐ 6. UB 1468 (sample 6172): Stenstorp–Dala, zone Vb.
Anterior side of specimen with very small, narrow lateral projections.

8–15: *Westergaardodina microdentata* Zhang 1983, p. 47, all ×130.

☐ 8, 11. UB 1469 (sample 5672): S. Möckleby, zone Vb.
Anterior and anterolateral view.

☐ 9. UB 1470 (sample 5672): S. Möckleby, zone Vb.
Anterolateral view.

☐ 10, 12. UB 1471 (sample 6313): Tomten, zone Vc.
Posterior and basal view. Note the bowl-shaped interior region.

☐ 13, 14. UB 1472 (sample 6313): Tomten, zone Vc.
Oblique views of a large specimen from both posterior and lateral sides. The proportional differences result from a perspective shortening.

☐ 15. UB 1473 (sample 6369): Brattefors, zone V undiff.
Anterior view.

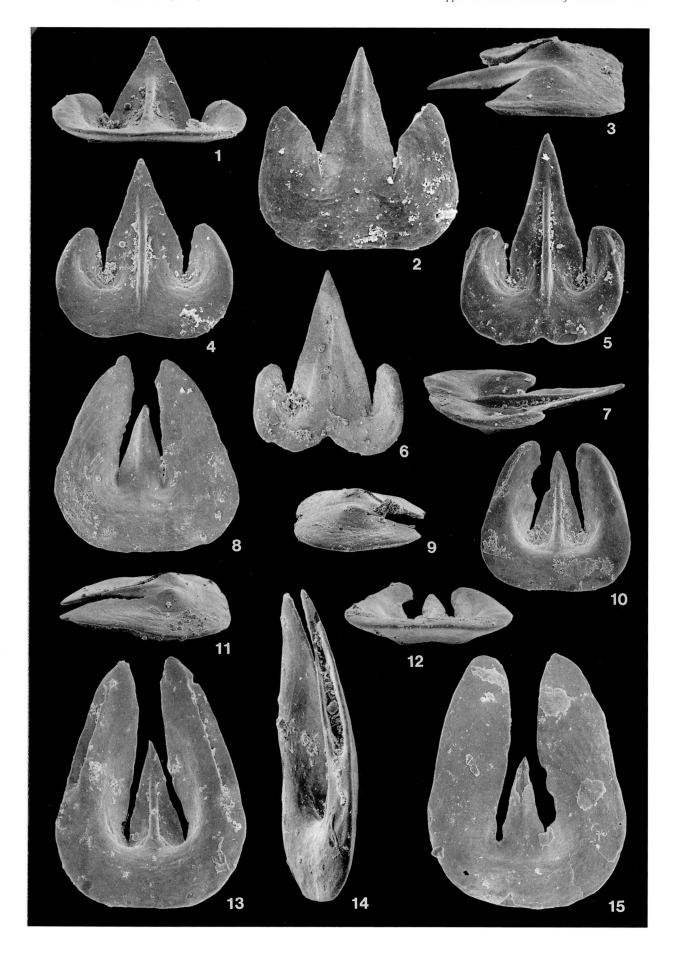

Plate 41

1–15. *Coelocerodontus bicostatus* van Wamel 1974, p. 53. Morphotype alpha.

☐ 1, 6. UB 1474 (sample 5968): Trolmen, zone V undiff.
   1. Lateral view; ×130.
   6. Basal view; ×180.

☐ 2, 7. UB 1475 (sample 5969): Trolmen, zone Vc.
   2. Lateral view; ×130.
   7. Basal view; ×180.

☐ 3, 8. UB 1476 (sample 7225): S. Möckleby–Degerhamn, zone Vc.
   3. Lateral view; ×130.
   8. Basal view; ×180.

☐ 4, 9. UB 1477 (sample 988): Trolmen, zone Vc.
   4. Lateral view; ×130.
   9. Oblique basal view; ×180.

☐ 5, 12. UB 1478 (sample 5658): Degerhamn, zone Vc.
   5. Oblique anterior view; ×130.
   12. Lateral view; ×130.

☐ 10, 13. UB 1479 (sample 6421): Trolmen, zone Vc.
   10. Lateral view; ×130.
   13. Basal view; ×180.

☐ 11, 14. UB 1480 (sample 6727): Trolmen, zone Vc.
Cluster of six sclerites, possibly coprolithic.
   11. ×260.
   14. ×130.

☐ 15. UB 1481 (sample 5669): S. Möckleby, zone Vc; ×130.
Lateral view.

16–21. *Coelocerodontus bicostatus* (van Wamel 1974), p. 53. Morphotype beta.

☐ 16. UB 1482 (sample 7232): S. Möckleby–Degerhamn, zone Vc; ×130.
Coprolithic cluster of at least five sclerites.

☐ 17, 19 UB 1483 (sample 5969): Trolmen, zone Vc.
   17. Lateral view; ×130.
   19. Basal view; ×180.

☐ 18, 21. UB 1484 (sample 5969): Trolmen, zone Vc.
   18. Lateral view; ×130.
   21. Basal view; ×180.

☐ 20. UB 1485 (sample 6727): Trolmen, zone Vc; ×130.
Lateral view.

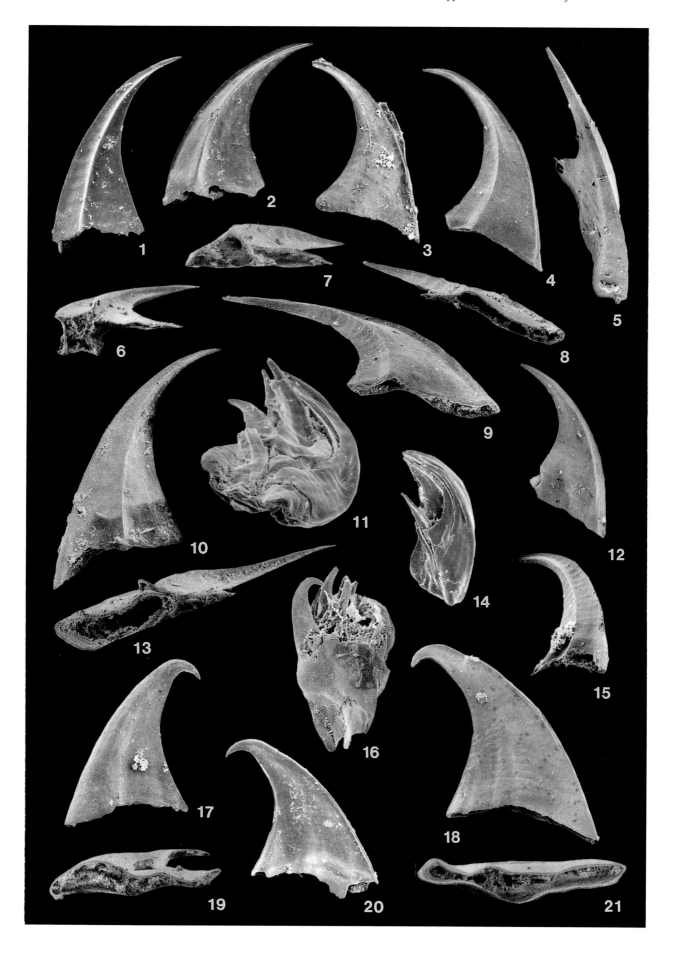

Plate 42

1–16. *Proconodontus muelleri* Miller 1969, p. 56.

☐ 1. UB 1486 (sample 6727): Trolmen, zone Vc.
Posterolateral view.

☐ 2. UB 1487 (sample 6881): Trolmen, zone V undiff.
Lateral view.

☐ 3. UB 1488 (sample 6727): Trolmen, zone Vc.
Lateral view.

☐ 4. UB 1489 (sample 990): Trolmen, zone Vc.
Lateral view.

☐ 5. UB 1490 (sample 6419): Trolmen, zone Vc.
Posterolateral view.

☐ 6, 7. UB 1491 (sample 6421): Trolmen, zone Vc.
Lateral and oblique view from above.

☐ 8. UB 1492 (sample 6421): Trolmen, zone Vc.
Lateral view.

☐ 9. UB 1493 (sample 990): Trolmen, zone Vc.
Lateral view.

☐ 10. UB 1494 (sample 6421): Trolmen, zone Vc.
Posterolateral view.

☐ 11. UB 1495 (sample 6419): Trolmen, zone Vc.
Posterolateral view.

☐ 12, 14. UB 1496 (sample 5968): Trolmen, zone V undiff.
View onto the tip and posterolateral view.

☐ 13, 16. UB 1497 (sample 6421): Trolmen, zone Vc.
Basal and lateral view.

☐ 15. UB 1466 (sample 5968): Trolmen, zone V undiff.
Lateral view.

17–21. *Proconodontus serratus* Miller 1969, p. 56.

☐ 17. UB 1498 (sample 6422): Trolmen, zone Vc.
Lateral view.

☐ 18, 19. UB 1499 (sample 6422): Trolmen, zone Vc.
Lateral view and detail of posterior keel.

☐ 20. UB 1500 (sample 6421): Trolmen, zone V undiff.
Oblique lateral view.

☐ 21. UB 1501 (sample 6422): Trolmen, zone Vc.
Oblique lateral view.

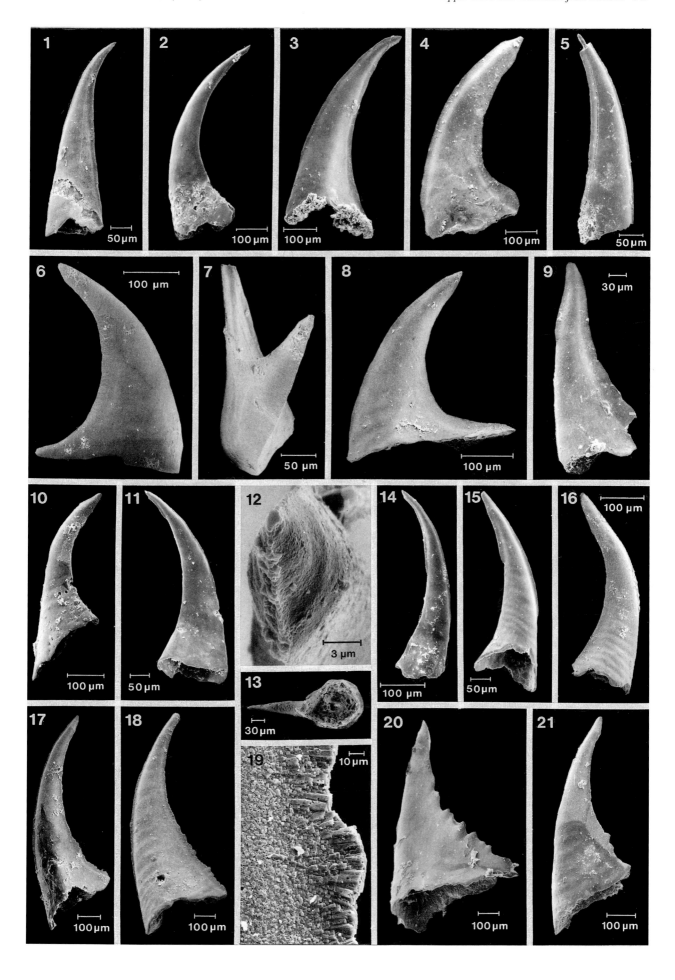

Plate 43

1–3, 6, 7. *Proconodontus serratus* Miller 1969, p. 56.

☐ 1. UB 1502 (sample 6421): Trolmen, zone Vc.
  Posterolateral view.

☐ 2, 6. UB 1503 (sample 6422): Trolmen, zone Vc.
  Lateral view and view onto broken tip.

☐ 3, 7. UB 1504 (sample 6421): Trolmen, zone Vc.
  Lateral and basal view.

4, 5, 8–15. *Cordylodus primitivus* Bagnoli, Barnes & Stevens 1987, p. 54.

☐ 4. UB 1506 (sample 6420): Trolmen, zone Vc.
  Lateral view.

☐ 5, 9. UB 1507 (sample 6420): Trolmen, zone Vc.
  Lateral view and view onto broken tip.

☐ 8. UB 1508 (sample 6422): Trolmen, zone Vc.
  Lateral view.

☐ 10. UB 1510 (sample 6421): Trolmen, zone Vc.
  Oblique lateral view.

☐ 11–13. UB 1511 (sample 7230): S. Möckleby–Degerhamn, zone Vc.
  Lateral and basal view.

☐ 14. UB 1512 (sample 6420): Trolmen, zone Vc.
  Lateral view.

☐ 15. UB 1513 (sample 7230): S. Möckleby–Degerhamn, zone Vc.
  Lateral view.

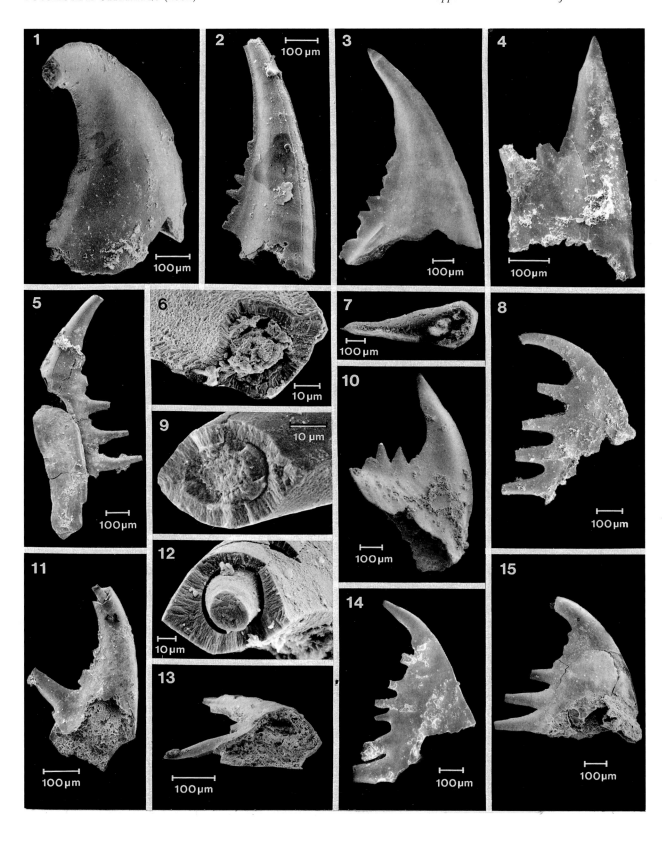

Plate 44

1–21. *Cambropustula kinnekullensis* n.gen., n.sp., p. 52, all ×90.

☐ 1. UB 1514 (sample 6784): Gum, zone I. Morphotype alpha.
Posterior view.

☐ 2. UB 1515 (sample 6414): Gum, zone I. Morphotype alpha.
Posterior view.

☐ 3. UB 1516 (sample 6750): Gum, zone I. Morphotype alpha.
Posterior view.

☐ 4, 5. UB 1517 (sample 6761): Gum, zone I. Morphotype alpha.
Anterior view from above and upper view.

☐ 6. UB 1518 (sample 6736): Backeborg, zone I. Morphotype alpha.
Posterior view.

☐ 7. UB 1519 (sample 6763): Gum, zone I. Morphotype alpha.
Posterior view.

☐ 8. UB 1520 (sample 6761): Gum, zone I. Morphotype alpha.
Posterior view.

☐ 9, 12. UB 1521 (sample 6760): Gum, zone I. Morphotype gamma.
Posterior and basal view.

☐ 10. UB 1522 (sample 6404): Gum, zone I. Morphotype alpha.
Posterior view.

☐ 11, 14. UB 1523 (sample 6776): Gum, zone I. Morphotype beta.
Posterior and basal view.

☐ 13. UB 1524 (sample 6763): Gum, zone I. Morphotype delta.
View from above.

☐ 15. UB 1525 (sample 5952): Backeborg, zone I. Morphotype delta.
Posterior view of specimen fotographed upside down.

☐ 16. UB 1526 (sample 6404): Gum, zone I. Morphotype alpha.
Basal view displaying growth lamellae.

☐ 17, 19. UB 1527 (sample 6736): Backeborg, zone I. Morphotype delta.
Posterolateral view and view from above.

☐ 18, 21. UB 1528 (sample 6751): Gum, zone I. Morphotype gamma.
Posterior and basal view.

☐ 20. UB 1529: Gum, zone I. Morphotype beta.
Oblique posterior view.

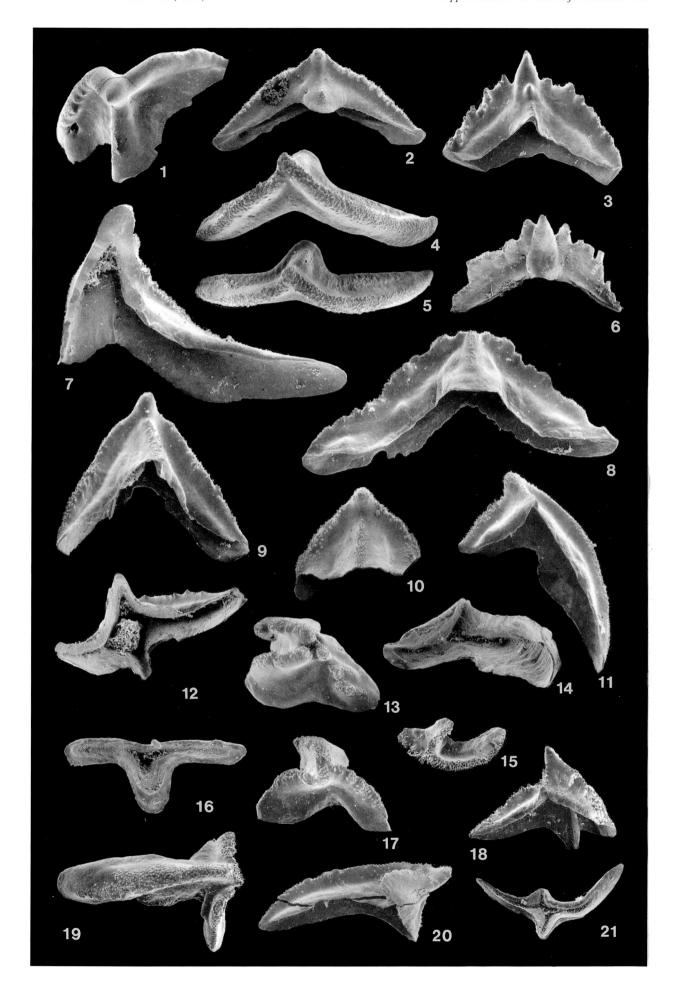

Plate 45

1–21. *Cambropustula kinnekullensis* n.gen., n.sp., p. 52, all ×90.

☐ 1, 4. UB 1530 (sample 6761): Gum, zone I. Morphotype gamma.
Anterior view and view from above.

☐ 2. UB 1531 (sample 6414): Gum, zone I. Morphotype gamma.
View from above.

☐ 3, 6. UB 1532 (sample 6761): Gum, zone I. Morphotype alpha.
Anterior and oblique posterior view.

☐ 5. UB 1533 (sample 6409): Gum, zone I. Morphotype gamma.
Posterior view.

☐ 7. UB 1534 (sample 6762): Gum, zone I. Morphotype gamma.
View from above.

☐ 8. UB 1535 (sample 6414): Gum, zone I. Morphotype gamma.
Posterolateral view.

☐ 9. UB 1536 (sample 6409): Gum, zone I. Morphotype gamma.
Anterior view.

☐ 10. UB 1537 (sample 6763): Gum, zone I. Morphotype gamma.
Anterior view from above.

☐ 11, 14. UB 1538 (sample 6776): Gum, zone I. Morphotype gamma. Holotype.
Oblique basal view and view from above.

☐ 12. UB 1539 (sample 6783): Gum, zone I. Morphotype gamma.
Posterior view.

☐ 13. UB 1540 (sample 6414): Gum, zone I. Morphotype alpha.
View from above.

☐ 15, 18. UB 1541 (sample 6763): Gum, zone I. Morphotype delta.
Anterior and posterior view.

☐ 16, 19. UB 1542 (sample 6761): Gum, zone I. Morphotype delta.
View from above and oblique posterior view.

☐ 17, 20. UB 1543 (sample 6424): Gum, zone I. Morphotype delta.
Posterior view and view from above.

☐ 21. UB 1544 (sample 5951): Gum, zone I. Morphotype gamma.
View from above.

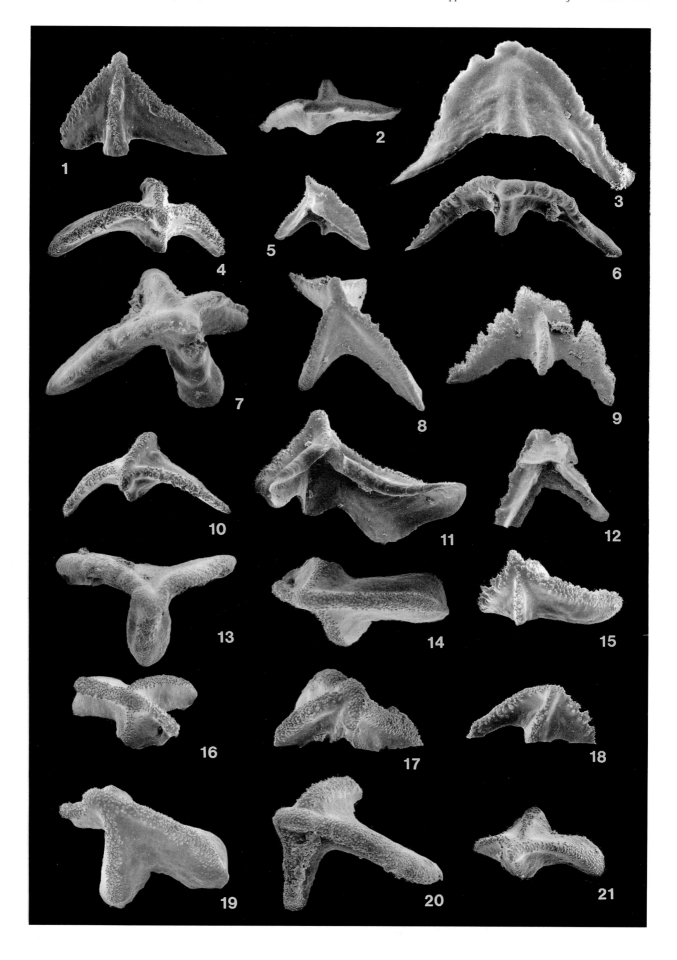